'At a time where architecture is
and Sussman remind us in *Cognitive Architecture* that buildings and cities profoundly affect our lives. Design impacts our senses, our psyche and our disposition, drawing on our history and our evolution as a species. The book brings us back to fundamentals, insisting that architectural design must consider and respond to our basic sensibilities. It returns architectural discourse to a place where the public at large can partake in the assessment of the built environment.'

Moshe Safdie, international architect and founder of Safdie Architects

'Buildings are intended to be viewed, traversed, and lived in by people, but theoreticians and practitioners of architecture have rarely thought deeply about human nature and why and how it needs to be taken seriously when designing buildings and urban areas. In this short book, one of a small but growing number of contributions that are rectifying this deficiency, Sussman and Hollander clearly describe why our attraction to edges, faces, symmetry, curves, and stories, all of which have deep evolutionary roots, should influence architecture and urban design. Every person who wishes to design buildings should read and internalize the message of this book before putting pen to paper.'

Gordon H. Orians, Professor Emeritus of Biology, University of Washington, Seattle, and author of Snakes, Sunrises, and Shakespeare: How Evolution Shapes Our Loves and Fears

'*Cognitive Architecture* is an important and insightful book that significantly links our responses to seemingly lifeless buildings to the very foundations of our inborn affinity for life. It takes our ancient intuitions about bringing buildings to life into the modern realm of rapidly unfolding discoveries in evolutionary biology and neuroscience. It critically goes beyond theory and science to the application of this knowledge to a better and more beneficial architecture and urban design.'

Stephen R. Kellert, Professor Emeritus and Senior Research Scholar, Yale University, Connecticut, and author of Birthright: People and Nature in the Modern World

'*Cognitive Architecture* is an inspiring book which introduces the reader to our sense of aesthetics that is at the core of human cognition in which we shape our built environment. This book will be of great value to architects, environmentalists, designers, geographers and planners, as well as students of those fields.'

Andreas Luescher, a Swiss architect and Professor of Architecture and Environmental Design, Bowling Green State University, Ohio

Cognitive Architecture

In *Cognitive Architecture*, the authors review new findings in psychology and neuroscience to help architects and planners better understand their clients for the sophisticated mammals they are, arriving in the world with built-in responses to the environment that have evolved over millennia.

The book outlines four main principles—Edges Matter, the fact that people are a thigmotactic or a 'wall-hugging' species; Patterns Matter, how we are visually oriented; Shapes Carry Weight, explains how our preference for bilateral symmetrical forms is biological; and finally; Storytelling is Key looks at our narrative proclivities that define our species and play a major role in successful place-making. The book takes an inside-out approach to design, arguing that the more we understand human behavior, the better we can design for it. The text suggests new ways to analyze current designs before they are built, allowing the designer to anticipate a user's future experience. More than one hundred photographs and drawings illustrate its concepts. Six exercises and additional case studies suggest particular topics—from the significance of face-processing in the human brain to our fascination with fractals—for further study.

Ann Sussman, AIA, an architect, works as an artist, writer and community organizer. With Justin B. Hollander, she created the *Open Neighborhood Project*, using art, craft, and digital tools to increase public participation in planning, and earning the Gold Star Award from the Commonwealth of Massachusetts in 2010. Her studio is at ArtScape in the Bradford Mill, an art and business enclave in Concord, Massachusetts, USA.

Justin B. Hollander, PhD, AICP, is an Associate Professor in the Department of Urban and Environmental Policy and Planning at Tufts University, Medford, Massachusetts, USA, and the author of three previous books on city planning and design: *Polluted and Dangerous* (2009), *Principles of Brownfield Regeneration* (2010), and *Sunburnt Cities* (2011).

Cognitive Architecture

Designing for How We Respond to the Built Environment

ANN SUSSMAN AND JUSTIN B. HOLLANDER

NEW YORK AND LONDON

First published 2015
by Routledge
711 Third Avenue, New York, NY 10017

and by Routledge
2 Park Square, Milton Park, Abingdon, Oxon OX14 4RN

Routledge is an imprint of the Taylor & Francis Group, an informa business

Library of Congress Cataloging in Publication Data
Sussman, Ann, author.
Cognitive architecture : designing for how we respond to the built environment / Ann Sussman and Justin Hollander.
pages cm
1. Architecture—Psychological aspects. 2. Space perception—Physiological aspects. I. Hollander, Justin B., author. II. Title.
NA2540.S87 2015
720.1'9—dc23
2014012617

ISBN: 978-0-415-72468-5 (hbk)
ISBN: 978-0-415-72469-2 (pbk)
ISBN: 978-1-315-85696-4 (ebk)

Acquisition Editor: Wendy Fuller
Editorial Assistant: Grace Harrison
Production Editor: Siobhán Greaney

Typeset in Didot and Helvetica
by FiSH Books Ltd, Enfield

To Rhoda Rappaport, professor emeritus of history at Vassar College (1961–2000). Professor Rappaport imbued her students with a passion for the History of Science, challenging them to recognize the importance of unconventional thinking and how it can promote cultural transformation.

Contents

List of Figures and Tables

Figures

Tables

1

A New Foundation: Darwin, Biology, and Cognitive Science

> Humans are not proud of their ancestors, and rarely invite them round to dinner.
>
> Douglas Adams

Evolution. It is the first and last word in this book. The thesis of this book is that the more we understand how human beings are an artifact of Darwin's theory of evolution, the more creatively and successfully we will be able to design and plan for them. Evolution holds that all life evolved, or transformed, from a common ancestor. This holds true for modern humans, a relatively recent species believed to have been on earth about two hundred thousand years. As such, we carry significant baggage from a very long journey: our planet Earth is 4.6 billion years old and the first life appeared on it some 3.6 billion years ago. In this book, we explore how our evolutionary path can be seen at work in the ways humans function, including how we walk, think, see, and prioritize viewing things in our environment. Our sense of aesthetics is at root biological, evolving over millennia.

Many books on architecture and planning refer to nature, but most do not talk about humans as evolved mammals with their perceptual systems a product of 'natural selection,'[1] the mechanism for evolution naturalist Charles Darwin defined in his most famous text, *On the Origin of Species by Means of Natural*

Selection in 1859. This book has this idea at its core. We believe that because of the burgeoning research on the brain and cognitive science particularly happening now in the early twenty-first century, more books like this will follow. Hardly a day goes by without some new finding in evolutionary biology, psychology, neuroscience, or genetics, reframing our understanding of what it means to be humans and how we came to be.[2]

There is a central paradox to architecture and planning that this book also addresses. Practitioners rarely meet the people who will be most affected by their work. Most buildings outlive their creators. Post-occupancy evaluations are expensive and infrequent. Even in residential design, with an average American staying in a house only 13 years, the building will likely long outlast its original tenants (Emrath 2009). What should the architect or planner know about the human as a generic client? How should they think about something as complex as 'human nature,' or establish guidelines for designing successful places for people never met?

The intent of this book is to answer these questions or at least provide a basic framework for doing so, teasing out innate human responses and expectations of the man-made built world. However, here again, we run into another central paradox that provides a foundation for this book. "Our perceptual systems are designed to register aspects of the external world that were important to our survival..." wrote Harvard psychologist Steven Pinker in *The Blank Slate: The Modern Denial of Human Nature* (2003: 199). In other words, we see the 'reality' nature intends us to see, the one that led to our species' survival in the past, which was for almost all of human history, the one outside in the natural world, and not man-made. Our oldest cities are only about 6,000 years old. We never evolved to live in the situations most of us find ourselves in today. What this suggests for architecture and planning is our subject.

Twentieth-century urban observers including the writer Jane Jacobs maintained the way forward in planning and architecture would be to better understand how people are "a part of nature." Jacobs, an outspoken critic of most mid-twentieth-century planning projects, criticized planners for treating people like cars. It certainly is easier to design for cars than people, she noted, and though it may be easy to treat people as though they were purely mechanical, proceeding this way never produces places with lasting public resonance. In *The Death and Life of Great American Cities,* she wrote:

> Underlying the city planners' deep disrespect for their subject matter... lies a long-established misconception about the relationship of cities—and indeed of men—with the rest of nature... Human beings are, of course, a part of nature, as much so as grizzly bears or bees or whales or sorghum cane.
>
> (Jacobs 1961: 443)

In the chapters ahead we outline what it means, according to our best interpretations of recent science, to consider people exactly as Jacobs would have it, as "a part of nature, as much so as grizzly bears or... sorghum cane." It turns out people have multiple subconscious tendencies and behaviors that govern their responses to built environments—no wonder they flummoxed mid-century planners. Jacobs was right: the planners' work overlooked essential aspects of human make-up. But, to be fair, these traits were not well documented at the time and, because they are subconscious, by definition can be hard to see. In the five chapters that follow we outline what these hidden aspects in human nature are, each in its own chapter, culling out the behaviors that we believe are most significant for architecture and urban planning.

Chapter 2, Edges Matter, begins with exploring a phenomenon that Jacobs found curious: people avoid the center of open spaces and tend to stick to the sides of streets, even in car-free zones. Understanding the biological and evolutionary basis of this subconscious tendency is critical for urban planning, particularly if planners hope to create walkable places. We discuss the recent psychological research describing the trait as a survival and orientation strategy and introduce its scientific name: *thigmotaxis*. We also look at thigmotaxis to demonstrate one of Darwin's essential insights: that nature is 'conservative,' or traits that are successful reappear in new species again and again. We chart the research record on thigmotaxis to help us appreciate how fantastically conservative nature gets. Researchers have documented thigmotaxis in organisms 3.6 billion years old and a host of other species that have evolved over millions of years since, including *Homo sapiens*, who are the relative newcomers on the planet.

Chapter 3, Patterns Matter looks at how the human brain does not treat our senses equally. Lacking the sonar of bats or the smelling skills of bears,[3] the human brain is essentially oriented toward vision. More of our gray matter is devoted to creating our visual representation of the world than anything else. The implications of this fact for design and architecture are significant, suggesting how important detail and visual diversity are for building elevations and urban layout. Once you realize that half the sensory information going to the human brain concerns visual processing (Kandel 2012: 238), no other conclusion becomes tenable. Moreover, our mental apparatus does not handle visual inputs equally either—it prioritizes the face. The evolutionary reasons for this are clear: identifying faces quickly, whether friend or foe, proved critical for survival. An apparent by-product of our finely evolved adeptness at face-processing is that we see faces everywhere and subconsciously arrange facial features out of random data. This includes, for instance, 'seeing' faces in inanimate things where they are not, from the 'man in the moon' to a 'Virgin Mary' in a burnt piece of toast (BBC News 2004). The impact of this trait on architecture and aesthetics is something we are only beginning to appreciate. Research suggests we also see faces in many of our favorite houses and streetscapes and in so doing most easily and subconsciously make emotional attachments to these places. We review computer science literature that suggests this tendency needs to be programmed into future robots to make them more human-like, as well as more capable helpmates able to anticipate our responses to our surroundings.

Further delving into the dominant characteristic of the face, in Chapter 4, Shapes Carry Weight, we consider bilateral symmetry, which is common to animals generally. People are bilaterally symmetrical and so is much we intentionally make, including the patterns in our craft and in many of our building and city designs (see Figure 1.1). We look at why this form prevails, again from a Darwinian perspective, learning that animals and humans subconsciously associate the shape with power and robustness. Bilateral symmetry carries deep, innate psychological significance. In research studies when human subjects of both sexes were asked to choose between symmetric and asymmetric faces—and even symmetric and asymmetric geometric patterns—they consistently preferred the more symmetrical. Looking at these studies is seeing natural selection at work.

While the psychological traits above we share with other life, no other creature has the one discussed in Chapter 5, Storytelling is Key: our narrative ability. This characteristic distinguishes humans and is the consequence of possessing complex neural circuitry unlike that of any other animal on the planet. Our brain size coupled with this story-enabling capacity contributes to making the human brain the outlier it is in the animal world. Our innate ability to invent stories and create multiple scenarios in any situation—and not necessarily act upon them—is considered highly adaptive. Our narrative proclivities, in turn, have led to the creation of new artifacts on earth: they make literature and art possible for one, and perhaps more significantly give people identity and a sense of meaning. As the narrative-telling species, we also are a passionate narrative-seeking one. Much as a horse favors an open field

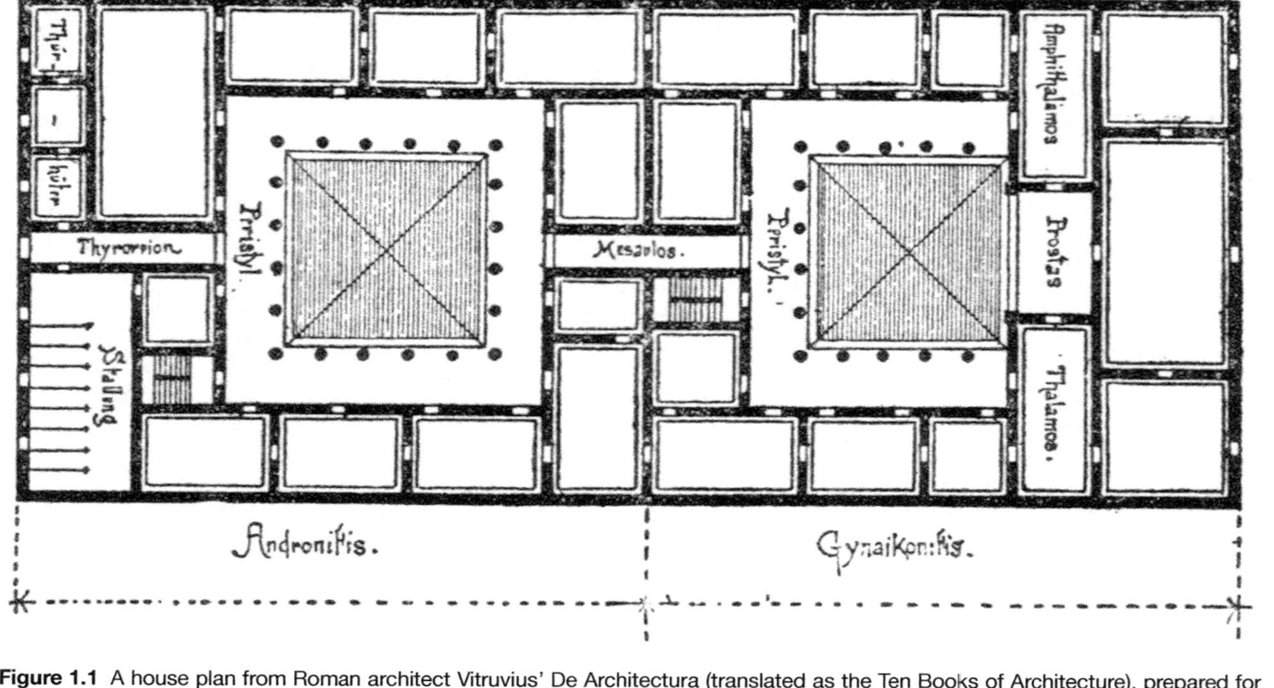

Figure 1.1 A house plan from Roman architect Vitruvius' De Architectura (translated as the Ten Books of Architecture), prepared for the Emperor Augustus, first century BCE. The plan features nearly symmetrical men's and women's sections ('men's quarters' [left]; 'women's quarters' [right]) and in each of these areas multiple bilaterally symmetrical elements including room plans and column layouts (Source: Wikimedia Commons).

where it can gallop, a beaver the tree-lined stream where it can build, people love settings that engage their storytelling behavior. This can be seen in the broad popularity of books, movies, TV, Netflix, or YouTube, and when visitors travel to far-flung places around the world to take in significant and unusual histories, in some ways it does not seem to much matter whether the stories tied to place are real or fanciful. Disneyland in Anaheim, California, discussed in Chapter 2, remains one of the most visited places on earth, engaging more than 650 million visitors (or more than twice the population of the United States) since opening in 1955 with multiple made-up narratives in an obviously staged setting. Many of the world's most famous buildings, sites, and cities also possess 'embedded narrative.' They contain a specific formal sequence in their design that, like a story, has a beginning, middle, and end, or similar sequence. Or, as we see when we look at the popular Italian Renaissance garden, Villa Lante, at the end of Chapter 5, a familiar biblical story used to determine the actual layout of the plan. Humans at once inhabit a physical realm and one that is bounded by narrative and quite immaterial. And like other animals, we favor those places and activities that most enable our species-specific abilities.

People also like looking at nature and seeing landscapes that recall their species-specific past. This is the subject of Chapter 6, Nature is our Context, which concludes and draws together the book's earlier sections. An expression of how we are 'a part of nature' as Jacobs noted, is that we love looking at life. The eminent Harvard biologist E. O. Wilson labeled the idea that there is an innate bond between humans and all other living things, "the biophilia hypothesis," in his book *Biophilia*. Our "urge to affiliate with other forms of life is to some degree innate," he wrote (Wilson 1984: 85). Our hunter-gatherer ancestors came into their humanness in the grassy plains with scattered trees of the African, and later European and Asian, savanna, Wilson explains. In a certain sense, no matter where we go today, tens of thousands of years later, we never leave this landscape behind. Given the possibility of living anywhere, people still "gravitate statistically" toward a savanna-like view, he noted, and "will pay enormous prices to have (it)" (Wilson, Chapter 2: The Nature of Human Nature, in *Biophilic Design*, by Kellert *et al.*: 23).

Biophilic design, the approach to building design rooted in the biophilia hypothesis, strives to ensure new projects recognize and meet the human need to observe and engage with nature. In this book we hope to tease apart the evolutionary scrim that humans look through to empower designers to not only make their projects biophilic, but also more ably anticipate and fit our humanness. The more you know about human behavior, the better you can design for it. We review the hidden tendencies at work that have been outlined in the previous pages here and suggest their implications for architectural and urban planning theory, policy, and practice. We do not expect readers to have prior knowledge of workings of the brain or its parts, but in the Appendix we outline different ways of thinking about its organization and its development.

A note on the title: We use *Cognitive Architecture* to explore how research in psychology and the cognitive sciences can inform our understanding of the impact of buildings and city design on people. We recognize that the term is also used in computer science and in cognitive science, referring to the basic design of computers in the first instance, and the information-processing organization of the brain in the second, but that is not our context.

Exercise for Chapter 1: Watching People

- Visit a suburban or urban center and, with camera, notebook, or mobile device in hand, observe people: where do they gather; how do they walk? Preferably draw or sketch; you tend to look at things more closely then. Where is there a specifically interesting and definable response to the built environment? This is in preparation for Chapter 2.

Notes

1 How does natural selection work? It is the mechanism for explaining how transformation happens and why so many diverse species exist. Charles Darwin explains it in the introduction to *Origin of Species* this way: "As many more individuals of each species are born than can possibly survive; and as, consequently, there is a frequently recurring struggle for existence, it follows that any being, if it vary however slightly in any manner profitable to itself, under the complex and sometimes varying conditions of life, will have a better chance of surviving, and thus be naturally selected. From the strong principle of inheritance, any selected variety will tend to propagate its new and modified form." Another term for the process is "survival of the fittest," which the English polymath Herbert Spencer later coined, and Darwin called "more accurate and sometimes equally convenient."

2 The many new books on the brain and how it influences our actions and decision-making include more than 100 titles exploring what recent research implies for pedagogy, psychology, and sociology (exclusive of medical texts). These include best sellers: Charles Duhigg's, *The Power of Habit: Why We Do What We Do in Life and Business*, Daniel Kahneman's, *Thinking Fast and Slow*, Eric Kandel's *The Age of Insight: The Quest to Understand the Unconscious in Art, Mind, and Brain, from Vienna 1900 to the Present*, and *Subliminal: How Your Unconscious Mind Rules your Behavior*, by Leonard Mlodinow.

3 Bears have been known to smell prey up to 40 miles away; bats can 'see' in the dark using echolocation, where they emit high-pitched sounds and interpret the echo that bounces back to determine object location.

2

Edges Matter: Thigmotaxis (the 'Wall-hugging' Trait)

> ... our streets no longer work. Streets are an obsolete notion. There ought not to be such things as streets; we have to create something that will replace them.
>
> Le Corbusier (de Botton 2006: 243)

> The needs of city planning (require) ... doing away with the corridor street.
>
> Preparatory Congress for Modern Architecture, Vaud, Switzerland, 1928

People, like many animals, tend to behave differently in a room inside than when outdoors. Outside in built environments, they seem more at ease when buildings create a room-like condition that surrounds them on several sides, like in the picture on page 11 of Piazza del Campo, the historic square in Siena, Italy (see Figure 2.1). Here residents and tourists gravitate to the sides of the piazza's medieval walls as though pulled by a magnet. "We appreciate buildings which form continuous lines around us and make us feel as safe in the open air as we do in a room," wrote architectural critic Alain de Botton (2006: 245) in *The Architecture of Happiness*. In this chapter, we explore why. We review what famous urban-observers saw when they watched people in cities and then look at more recent scientific findings to help make sense of their findings.

Figure 2.1 Tourists and residents gather at the edge of Piazza del Campo, Siena, Italy (Source: Garry D. Harley).

A Great Place for People-watching

555 Hudson Street in Lower Manhattan is a three-story brick building in Greenwich Village that is fairly nondescript (see Figure 2.2). Some 5 meters wide (about 16 feet), it was built circa 1901, with a storefront on the first floor and living quarters above and measures out at close to 195 square meters (2,100 square feet). The Hudson River is three blocks and less than three hundred meters (a quarter mile) away. It turns out, though, that 555 Hudson and its surrounding neighborhood of low-rise

Figure 2.2 On the left, 555 Hudson Street, Jane Jacob's narrow New York City town-house in the West Village, Lower Manhattan (partially hidden by scaffolding from a neighboring building) (Source: Ann Sussman).

storefronts, is an excellent vantage point for at least one thing: people-watching. That is what writer and urban observer Jane Jacobs did after she and her husband bought the building in 1947 for $7,000. At the time the area was a culturally diverse artists' haven and very affordable (in 2009, 555 Hudson sold for $3.3 million).

What Jacobs saw from 555 Hudson and its environs informed her most famous book, *The Death and Life of Great American Cities* (1961), which went on to become one of the most well-known titles in American urban history. The book promoted her efforts to push twentieth-century urban design and planning into a more human-centered direction, one where designers consider human behavior as well as accommodating the twentieth-century's proliferating numbers of new cars. Highway construction was booming in Jacobs' time, with automobile ownership soaring. A new highway was once planned not far from her own home. In part through her efforts to alert the public, the 10-lane Lower Manhattan Expressway that would have razed historic neighborhoods around 555 Hudson never happened.

One thing Jacobs realized is that people in outdoor urban spaces exhibit some unusual behaviors that typical urban plans and their originators do not take into account. No wonder the planners struggle to figure people out; car requirements really are more obvious. Pedestrians do not merely walk down the sidewalk, they perform an "intricate sidewalk ballet," she wrote, famously describing her own routine on Hudson Street:

> I make my own first entrance into it a little after eight when I put out my garbage can, surely a prosaic occupation, but I enjoy my part, my little clang, as the droves of junior high school students walk by the center of the stage dropping candy wrappers.
>
> (Jacobs 1961: 50–1)

This activity is meaningful:

> The ballet of the good city sidewalk never repeats itself from place to place, and in any one place is always replete with new improvisations.
>
> (Jacobs 1961: 50)

Significantly, Jacobs noted how the buildings and the way their first floor elevations are designed, seems to invite the dance. The buildings need to sit in a specific way. They "must be oriented to the street. They cannot turn their backs or blank sides on it and leave it blind" or the street will not generate much of a show. She writes about how buildings seem to exert an invisible pull. When children leave a child-care center in the city, for example, they are happier and feel safer walking home alongside buildings than by a wide-open city park (Jacobs 1961: 74). On pedestrian streets, even without cars around, people avoid the emptiness in a street center:

> They do not sally out in the middle and glory in being kings of the road at last. They stay to the sides.
>
> (Jacobs 1961: 374)

She observes this in Disneyland in Anaheim, California, in shopping malls, in places where cars are relegated to far away parking lots (p. 374). Even in the car-free realms, "people stay to the sides except where something interesting to see has been deliberately placed out in the "street" " (p. 374). She hypothesizes why:

> In more ordinary circumstances, people are attracted to the sides, I think, because that is where it is most interesting. As they walk, they occupy themselves with seeing—seeing in windows, seeing buildings, seeing each other.
>
> (p. 348)

Some fifteen years later, another keen urban watcher made careful note of 'edge-appeal,' too. Christopher Alexander, an architect and University of California at Berkeley professor, is much like Jacobs, a critic-at-large of twentieth-century development. In *A Pattern Language* (1977), he published meticulous observations of human behavior in a 1,150-page book exhaustively listing 253 'time-tested' patterns for guiding a project at all scales and sizes. Alexander cites "a primitive instinct at work," the human tendency to protect our back, as the reason people shun open spaces. In *Pattern 124, Activity Pockets*, he wrote:

> The life of a public square forms naturally around its edge. If the edge fails, the space never becomes lively.
>
> (Alexander *et al.* 1977: 600)

And again:

> People gravitate naturally towards the edge of public spaces. They do not linger out in the open.
>
> ...a big space will be wasted unless there are trees, monuments, seats, fountains—a place where people can protect their backs, as easily as they can around the edge.
>
> (p. 606)

Like Jacobs, he picked up on the importance of windows at street level:

> ...unless the building is oriented toward the outside, which surrounds it, as carefully and positively as toward its inside, the space around the building will be useless and blank—with the direct effect, in the long run, that the building will be socially isolated, because you have to cross a no-man's land to get to it.
>
> (p. 606)

In *The Pattern 160 Building Edge* (p. 753) he laments how in many modern buildings, "the space around it is not made for people" (p. 754).

And last but not least, even Le Corbusier, the Swiss architect who helped define modern architecture in the early twentieth century—and would do so much to take buildings entirely off corridor streets—could not help but marvel at the consistently robust appeal of sidewalks in the densely populated French capital in his book, *The Radiant City*, first published in 1922:

> In Paris I often walked through the district bounded by the Place des Vosges and the Stock Exchange—the worst district in the city and the most wretchedly overcrowded. Along the streets, on the skimpy sidewalks, the population moves in single file. By some miracle of group identification and the spirit of the city, even here people laugh and manage to get along, even here they tell jokes and have a good time, even here they make out.
>
> (Le Corbusier 1967: 12)[1]

Why do city streets generate this activity? How come people find streets magnetic? And why do they shun an open center?

It turns out continuous edges and street corridors aide and abet our movement, and this is an artifact of our evolution, as Alexander surmised. It is also a direct and underappreciated consequence of how humans are built, move, and where they tend to look. The Danish architect and planner Jan Gehl sums up the importance of architects and designers understanding how humans move in his book *Cities for People* (2010: 33). One of the most fundamental things to know about the human client, he says, is the way people naturally walk, which is summarized in the box opposite.

Because danger generally lurked on a horizontal plane for our ancestors, we have evolved eyes parallel with the ground the better to scan it. There is a natural tilt to the human head while

People are bipedal,
they have two feet,
and they walk with
eyes facing forward.
People rarely look
backwards or up.
They almost never
walk sideways or
backwards. They
dislike taking stairs.

walking of about 10 degrees to take in the path in front, which through the eons has persisted since it apparently kept us out of trouble (Gehl 2010: 39). Below we have diagrammed an 'Ambulation Man' and 'Ambulation Woman' walking in their natural stance with their heads slightly bent in this fashion (Figure 2.3).

Walking is something humans have done a very long time and our ancestors got good at. We are unique in this habit: no other mammal so successfully ambulates on two feet. The choice also proved fateful, enabling the development of our brain's size and specialized circuitry, which we consider later in the book. In the Paleolithic time, it has been estimated a woman walked an average of 9 miles a day; a man, 12.[2] At that rate they could cross the continental US in a year (Lieberman 2013). Ancient thinkers

Figure 2.3 'Ambulation Man' and 'Ambulation Woman' walk with their heads at a natural tilt of about 10 degrees (Source: Trey Kirk).

took note of the connection between walking and human health. "Walking is man's best medicine," Hippocrates, considered the father of western medicine, wrote in the fifth century BCE. (It is the lack of comparable exercise today that is considered to be a contributor to many modern health issues, including obesity and heart disease).

While stairs are an important tool in an architect's arsenal (and designing them successfully makes up part of the architectural licensing exam in the US), people usually avoid them whenever they can. We generally look to save energy if we can. Studies have shown that, given the choice of stairs or an escalator, people will pick the escalator 97% of the time. Reminded with a note that taking the stairs is actually a good way for modern humans to get exercise, people still pick the escalator 93% of the time (Lieberman 2013). People favor risk-free shortcuts, and tend to shun things that require conscious effort and paying extra attention. Humans also favor feeling safe and protected, particularly as they navigate through a new space. This is where our ancestral habits come into play, too. When it comes to edges, biologists classify humans, exactly like other mammals of prey, as thigmotactic, or a 'wall-hugging' species. In this chapter we review the literature that correlates the trait with levels of anxiety in people and our other animal relations. Thigmotaxis is what Jacobs and Alexander observed when they watched people avoiding wide-open spaces in California or New York City. If these dedicated students of urban behavior did not use the term, it may be because mid-twentieth-century scientists were researching thigmotaxis in mice, rats, and single-celled organisms—not people yet.

Thigmotaxis: A Hidden Trait

Thigmotaxis is effortless and instinctive. Many outside stimuli influence how we move and navigate without us having to stop and consciously think about it. Largely subconscious mental activities govern our behavior. Indeed, to put it most accurately, all of our conscious thoughts and actions are subconscious first, explains neuroscientist Eric Kandel, in *The Age of Insight* (Kandel 2012: 461–72). Sigmund Freud summed up the overwhelming mysteriousness of our cognitive processes, with a metaphor: "The mind is like an iceberg, it floats with one-seventh of its bulk above water." Governing our behavior from the unseen depths, thigmotaxis appears to be at work consistently as we move in our surroundings, although the metrics to consider and measure the trait in design do not yet seem to exist. (The hippocampus, an area of the limbic system, or paleomammallian part of the brain, is believed to govern navigation and spatial traits such as thigmotaxis; it is on top of the brainstem, buried below the cerebral cortex, the two hemispheres at the top of our head. For more on our brain morphology and function and how it compares with other animals', see the Appendix.)

Thigmotaxis is Old

As a primal way-finding strategy, biologists call thigmotaxis 'phylogenetically old' or old in terms of evolution. "The remarkable thing that Darwin discovered is that evolution is very conservative," wrote neuroscientist and Columbia University professor, Eric Kandel (2012). "If it finds through natural selection that some set of mechanisms work, it tends to retain those mechanisms in perpetuity." Table 2.1 illustrates just how seriously 'conservative' biologists believe life can get. Researchers have observed thigmotaxis in bacteria 3.6 billion years old, in amphibians and reptiles 300 million years old, and in us, the relative newcomers on earth, a young approximately 200,000 years old. Looking at the chart we learn that we share the tendency with sperm, the gene that is reportedly 600 million

Table 2.1 Evolutionary timeline for the 'wall-hugging' trait thigmotaxis with the date the tendency was first documented in the scientific literature by species at far right

Group	Example	Source
Bacteria (3.6 billion years ago)	*Paramecium aurelia*	Jennings, 1897
Early animals (600 million years ago)	Spermatoza	Dewitz, 1886 (cited in Jennings) Massart, 1888, 1889 (cited in Jennings)
Bilateria (550 million years ago)	Earthworm	Doolittle, 1971
Fish (500 million years ago)	Zebrafish	Schnörr *et al.*, 2012
Insects (400 million years ago)	Caterpillar Fruit fly (*Drosophila melanogaster*)	Steinbauer, 2009 Besson and Martin, 2004
Amphibians (360 million years ago)	Frogs	Bilbo, 2000
Reptiles (300 million years ago)	Snakes	Greene *et al.*, 2001
Mammals (200 million years ago)	Rats Humans	Barnett, 1963 Kallai *et al.*, 2007

Source: Devin Merullo

years old according to recent research at Northwestern University Medical School (2014). If sperm were not thigmotactic, come to think of it, we might not be here to observe the trait in ourselves.

The term thigmotaxis appears to have entered the scientific lexicon at the end of the nineteenth century. Scientists looking at single-cell organisms under microscopes noticed that some cells, single-cell paramecium, or sperm for instance, traveled along the edge of a solid introduced in the slide. The word, from ancient Greek, thigma- meaning to touch and -taxis meaning arrangement, has come to mean the direction of movement in response to an outside stimulus, or more simply, 'wall-hugging' (Schnörr *et al.* 2012: 37).

Studying thigmotaxis in mammals seems to have followed sometime after the creation of the first laboratory mazes to test animal behavior. The first such maze is believed to have been built at Clark University in Worcester, Massachusetts at the end of the nineteenth century. Very creatively, it was modeled after a human one, the famous seventeenth-century Hampton Court Maze, built as royal court entertainment in the outskirts of London and still apparently world-renowned for amusing tourists (see Figure 2.4) (Historical Royal Palaces 2014).

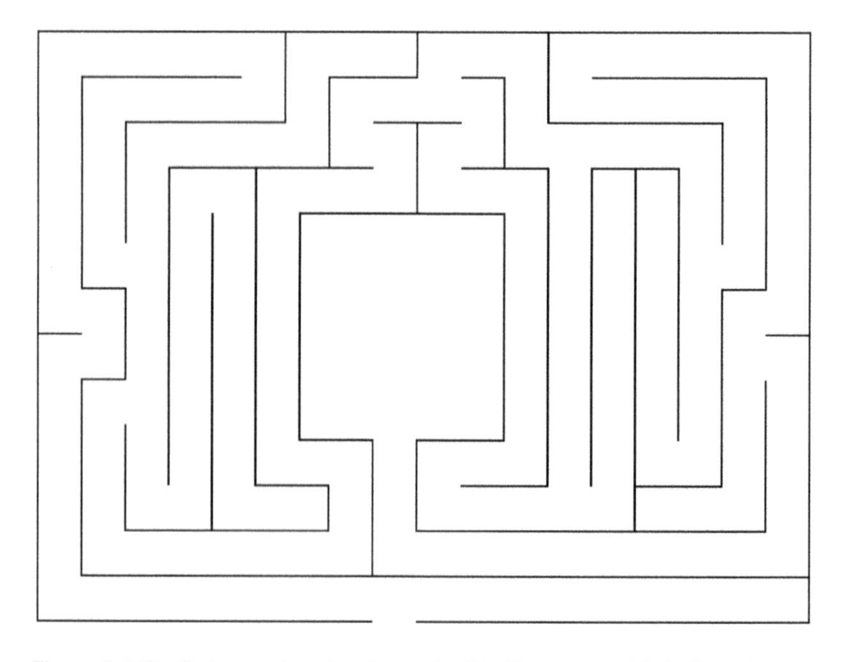

Figure 2.4 The first maze for laboratory animal testing was modeled after a human one designed for entertaining English royals; later versions were used in laboratories to study animal and human thigmotaxis (Source: Janice Ward).

Researching Thigmotaxis in Animals and People

Scientists have been watching rat and other lab animal behavior in mazes since, designing these in a variety of forms from T-shapes to radially shaped layouts, rather than human entertainment venues. Whatever pattern, they noticed the animals initially hugged the container sides and did not venture into its center. A recent study found that rats in a checkerboard maze where squares had one, two, or three walls, preferred the squares with the most sides. The early observers of mammalian maze behavior also note that thigmotactic behavior, which is relatively easy to follow and measure, increases with anxiety and decreases with familiarity: as rats knew the maze better they become less edge-oriented. Researchers describe this as an energy-conserving strategy and a survival one; animals that were more thigmotactic were more likely to survive, passing on the behavior in their genes to the next generation, through evolving species and eventually to us.

By the early 2000s studies to tease apart the specific role thigmotaxis plays in human navigation appeared in scientific publications. Not surprisingly, to analyze the 'wall-following' trait in people, psychologists built mazes for them. Sometimes these were real and sometimes virtual on computer screens (Kallai *et al.* 2007). This seems to be because humans can engage with their environment visually as well as by touch. In 2007, collaborating with psychologists at the University of Arizona and University of Southampton, psychologist Janos Kallai, at the University of Pecs, Hungary, sought to clarify the role thigmotaxis played when people move in a space:

> Although thigmotaxis is a well-characterized behavioral tactic commonly observed in nonhuman animals, its role in human navigation is yet to be explained... When an animal initially explores an enclosed place, it tends to stay in close contact with the perimeter of that space... One may quantify the tendency to

> avoid the inner zone of an open field by measuring either the time or the path length that an organism spends in close contact with the wall.
>
> (Kallai *et al.* 2007: 22)

Kallai and his fellow researchers knew that thigmotaxis was genetically based in humans like animals, and that it was one of several phases to human spatial learning, but when was it used? Were some people more prone to the trait than others? To find out the team selected 106 participants, men and women, chosen from respondents to a newspaper ad.

> We assessed a carefully selected sample of participants with different levels of fear and anxiety. Our purpose was to examine cognitive and emotional factors that may underpin thigmotaxis in virtual and physical arena mazes and in different spatial and non-spatial learning and memory tasks
>
> (Kallai *et al.* 2007: 23)

The participants took a battery of psychological tests, including ones that rated intelligence and general anxiety levels. Setting out to separate their cognitive functioning from emotional responses, the psychologists gave the subjects two timed trials, one where they viewed a PC screen; the other where their eyes were covered, they could not see and had to literally feel their way around. In the visual test, participants tried to locate a target in a circular arena on a computer navigating with a joystick. Their task, to find the target as quickly as possible, was timed and tracked. In the real maze, a circular wooden structure, 2 metres high and 6 meters in diameter (7 feet high; 21 feet in diameter), participants were fitted with opaque goggles, and led to a space that had eight objects in it, each shaped differently and set on a low stand. The target, a round object on the floor in one quadrant of the arena, emitted a tone when stepped on. Participants were instructed to find the noise-emitting target using the sculptural-cues as needed within 5 minutes or less. They too were timed and tracked.

Based on analysis of these findings and others from earlier studies, Kallai and co-authors concluded human thigmotaxis is similar to thigmotaxis in animals: "a tendency to refrain from exploring the inner zone of a novel place"; people have a bias to avoid centers and seek safety by sticking to the sides. As is true in all species, there is individual variation (another one of Darwin's major observations.) Not all people rely on thigmotaxis to the same degree. Just as human cognitive and anxiety levels differ, people who can create mental spatial maps more readily, do less wall-following. (Kallai *et al.* 2007: 27) And people who cannot form a mental image of a new space quickly, will linger at the side longer. These more anxious individuals, the authors contend, tend to stay on the wall until their minds have created 'a map' and they feel safer. The researchers conclude:

> Fear...triggers a specific exploratory strategy such as thigmotaxis, which plays an essential preparatory role in the first phase of spatial learning. The use of thigmotaxis helps the individual define the borders of an enclosed space and identify escape routes from that space. Thigmotaxis also provides the individual with the elements of an egocentric frame of reference...With the elements of that frame of reference in hand, the organism can begin to construct a cognitive map.
>
> (Kallai *et al.* 2007: 28)

In sum, thigmotaxis has several functions; initially, it is a preparatory strategy to help a person sense the borders of a space and its escape routes. It also helps us gather necessary data to locate ourselves in a specific place and from that 'home base' go on to construct a mental 'map' of the surroundings. People may be object-driven, too (running to the food truck in the center of a parking lot), they can use landmarks to direct their travel, such as a church steeple, mosque minaret, or turrets of a theme park castle, yet thigmotaxis remains a baseline strategy for navigation and initial exploration. We also scan our environment to understand it, but researchers believe 'wall-hugging' plays a key role here, too:

> Thigmotaxis defines the borders of space and visual scanning reframes it. We humans appear to use these strategies during our everyday activity in novel situations.
>
> (Kallai *et al.* 2005: 193)

What does this mean? For one, we navigate more like rats and mice than we may like to think. It can make us uncomfortable to think this way (see comment from English humorist Doug Adams at the top of Chapter 1). Yet, given that nature is 'conservative,' as the scientists describe, it fits. Second, the research confirms what Jacobs, Alexander, and others noted—well-defined street corridors appeal to us, and promote all kinds of other responses, the behaviors Jacobs poetically labeled "the ballet of the good city sidewalk." By contrast, when edge conditions are ill-defined, we instinctively go on alert. Like a train without a track, we have no way to engage, at least no easy way forward. Clear edge conditions, on the other hand, do so much: they can release us from anxiety, enable our subconscious construction of mental maps, suggest a way forward that fits our bipedal frame, and our preferred way of holding our head, all the while helping us conserve energy. After all, we can hope, but we can never quite predict what might be coming around the corner.

Understanding thigmotaxis can also help designers develop parameters for their work and anticipate its impact. As an example, Figure 2.5 shows a photograph of the Edward W. Brooke Courthouse in Boston, Massachusetts completed in 1999 (Kalmann McKinnell and Wood). It has tall, wide piers at its base that do not reference our human scale and impede views at our eye-level near the building's entrance. This makes it difficult for people to see the area in front of them as they enter or leave

Figure 2.5 The arcade of the Edward W Brooke Court House in Boston (1999, Kalmann McKinnell and Wood) does not acknowledge the way people like to walk, inadvertently promoting confusion in the urban landscape (Source: Ann Sussman).

the building. As a consequence, navigation becomes more energy intensive, since we really do prefer to walk with our eyes looking straightforward, with our head in a slight downward tilt, as Gehl notes. Perceiving this space, we think: what might be lurking behind those sharp-edged piers, anyway? Which way is best to walk? Is there a better path? Do I have to look up here? (which takes a bit more energy and I would prefer not to). The courthouse arcade demands a high level of user alertness and psychological arousal.

Contrast this space in Boston with the arcade in Figure 2.6 in Paris (Rue de Rivoli), near Place de la Concorde and the Louvre. The Parisian structure provides a reassuring walking path. It takes much of the uncertainty and guesswork out of walking in a large city. Everything seems so simple here. Although grand, the arcade acknowledges our scale. Orientation is easy; the layout directs you precisely where to go. The design acknowledges how we are built, anticipates the way we like to walk. It takes care of our needs and in the process, exalts them. You might imagine feeling like a king or at least a member of the lesser nobility, strolling down this walkway in Paris. The medieval street in Figure 2.7 in Siena, Italy, enables the same effortless stroll in a simpler fashion (note how the design aides and abets Ambulation Man's walk down the street). The environment and the person act as one.

Thigmotaxis at Work in Cities

Awareness of thigmotaxis can make pedestrian behavior much more understandable for today's city designers. Here is another example from central Boston. One of the most successful pedestrian areas in Boston is Hanover Street, an old road in the city's North End. Interestingly, it was laid out in pre-colonial times by indigenous peoples as a route to the harbor. Figure 2.8 is a figure-ground drawing of the road and its surrounding neighborhood

Figure 2.6 The eighteenth-century Rue de Rivoli arcade in Paris supports the ways that humans walk and encourages our movement forward. It was designed by Napoleon's architects (Source: Garry D. Harley).

Figure 2.7 A corridor street in Siena, Italy, invites you (and Ambulation Man) forward, encouraging an effortless stroll (Source: Garry D. Harley).

today: Jane Jacobs called out this same street as particularly vital fifty years ago (and in *The Death and Life of Great American Cities* admonished planners of her time for considering it a prospective target for demolition and urban renewal).

Hanover Street wends northward, narrow (10 meters [32 feet] or less) with very well-defined edges. As the figure-ground diagram shows, it is the only continuous street running from the bottom to the top of the drawing. Hanover Street is bordered by the wide open urban park, the Rose Kennedy Greenway to the south, and to the north, at the top right (and barely visible in the diagram) Boston Harbor. Hanover Street fits our bipedal design and forward-moving and second-guessing nature. It features two continuous side walls and a cross street encountered every minute or so, walking at a leisurely pace. Not surprisingly, most older cities from ancient ports such as Pompeii, to old-world

Figure 2.8 The figure-ground drawing of Boston's North End shows Hanover Street as the only continuous road running north to Boston Harbor, which is visible in the top-right corner (Source: Nora Shull).

capitals like Paris, were built this way: in an uncertain world, people wanted to make outside places feel more secure, not less. Like many other urban designs laid out in pre-automotive times, the North End features small blocks, abundant intersections, and many walking paths in and out (these can also be thought of as escape routes) (see Figure 3.28 for photo).

What may be most instructive about Hanover Street, however, is its location close to (some 400 meters [1,312 feet] away) a modernist urban renewal project from the 1960s which demolished more than a dozen urban blocks. This makes comparing our old building habits with newer ones relatively straightforward. Figure 2.9 shows what used to be near Hanover Street,

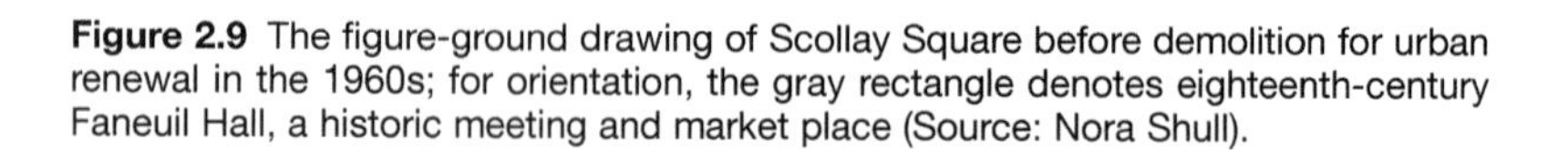

Figure 2.9 The figure-ground drawing of Scollay Square before demolition for urban renewal in the 1960s; for orientation, the gray rectangle denotes eighteenth-century Faneuil Hall, a historic meeting and market place (Source: Nora Shull).

c. 1829, before demolition. It was called Scollay Square, and was once a bustling neighborhood with short blocks and well-defined walking corridors. It is diagrammed in its current configuration, post-1960s urban-renewal-program, below (Figure 2.10). In the nineteenth and early years of the twentieth century Scollay Square was a haven for Irish immigrants fleeing

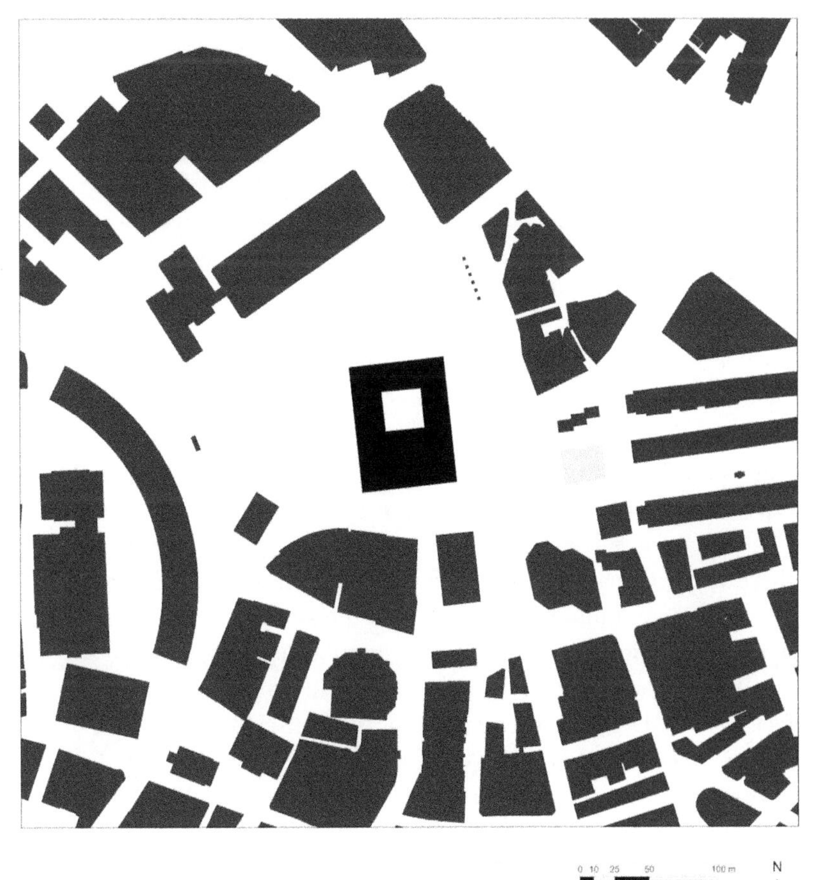

Figure 2.10 Figure-ground drawing of the Scollay Square area after urban renewal to create the new Boston City Hall (rectangle in center) and broad City Hall Plaza (Source: Nora Shull).

the Potato Famine (1850). It would flourish for decades as a busy entertainment and entrepreneurial zone with theaters, restaurants, residences, and in its latter decades, burlesque shows, as well as decidedly less savory enterprises. (Scollay Square also figures significantly in the history of science. Thomas Edison, the prolific American inventor, filed his first patent for a vote-counting machine developed in an attic lab here.) The gray rectangle, which is Scollay-Square adjacent, serving as a reference point in both diagrams, shows Faneuil Hall, the city's first market and meeting hall originally built in 1840. In the years following the Second World War, with residential flight to the suburbs, Scollay Square declined precipitously. By 1961, razing its 22 blocks was considered a swift way to promote central Boston's urban and financial renewal in a stroke. Remarkably, even in its decline, locals recall the district as a pedestrian magent. "My mother-in-law always felt safe walking through Scollay Square," says David Kruh, author of *Always Something Doing: Boston's Infamous Scollay Square* (1999).

Figure 2.10 shows Boston's Government Center, with City Hall and the wide City Hall Plaza today: Scollay Square's replacement would become a pedestrian void (for photo, see Figure 6.3). It represents an early modernist planning ideal: there are no small streets, blocks, or fussy intersections that slow traffic; no building front doors at the street. On the other hand, there does not seem to be any part of the redevelopment that takes into account how humans actually ambulate on two feet or use edges to navigate space or, subconsciously, are alert to their safety. The few super blocks offer pedestrians no hint of where to go, no suggestion of where to seek refuge. The City Hall building itself has no 'windows' on the street, as Jacobs and Alexander recommend. As a consequence, as Alexander's pattern language (Pattern #124) predicts, the urban renewal area became a 'no man's land.' In the half century since construction, Bostonians have consistently shunned the Plaza, and many do not think highly of the brutalist civic building at its center.

Repeated attempts at the Plaza's rehabilitation have met without measurable success to date. Ironically, Hanover Street and Boston City Hall plaza, so close on the map that they almost touch, could not be farther apart in terms of the public reaction they provoke. As such they provide a very useful example—a sort of living laboratory—of the human response to edge conditions. The plaza's urban renewal history illustrates the high cost we and future generations pay when planners and architects do not appreciate their clients as evolved mammals with embedded reactions to place.

What we confront in Boston's City Hall Plaza might be labeled 'a mismatch' between our ancient genes and our modern building habits. Psychologist Daniel Kahneman describes a similar sort of 'mismatch' in the modern diet in his popular book, *Thinking Fast and Slow* (2011). Our modern dietary practices, he argues, which make processed foods, sugars, and fats readily available, give our bodies many more of these nutrients than they can healthily handle. Similarly in architecture, the early modernist tendency to do away with streets and intersections in an effort to reduce congestion, or bring people closer to nature by putting high rises in park-like settings, ignores how people use buildings as orientation devices and protection screens. Ironically, modernism hampered the very populist values it intended to promote. Le Corbusier's plan for a 'Radiant City' of tall buildings was daring, but did not consider how building alignment encourages our movement and gets us to dance. We have learned in the decades since that we cannot do away with those 'corridor streets,' despite his pleas and the decrees of the 1928 Preparatory Congress for Modern Architecture, which are noted in this chapter's opening quotes, without erasing parts of ourselves and eons of our developmental history. Modern research is showing how much we are 'of nature;' we are tied to earth and life processes, including evolution.

Thigmotaxis Indoors

Humans rely on thigmotaxis navigating indoors, too. Given our edge-sensitivity, it becomes easier to understand why indoor shopping malls are *successful*, because they often duplicate the plan of an old-fashioned commercial street. They have stores on both sides or 'double-loaded corridors' in real-estate parlance. Perhaps from watching clients carefully, mall developers know what makes buyers happy and how aligning storefronts opposite each other keeps people busy and buying. Ancient bazaars in the Middle East use the same layout. The Grand Bazaar, or Kapalicarsi, (meaning 'Covered Bazaar') in Istanbul, Turkey, shown in Figure 2.11, was built as a series of double-loaded corridors six hundred years ago, under the Ottoman sultans and still thrives, as the name implies as a covered retail street.

Thigmotaxis Anchors 'Prospect and Refuge'

We suspect thigmotaxis is the trait moving us forward to seek and find 'refuge.' English geographer Jay Appleton proposed in his book, *The Experience of Landscape* (1975), the 'prospect-refuge' theory, which is familiar to many planners. The theory describes how people are drawn to edges to protect their backs, and also seek safe spots to take in broad landscape vistas. Thigmotaxis grounds the 'prospect-refuge' habit in evolutionary biology (see Table 2.1). Not only triggered when we take in vistas outdoors, as mentioned above, thigmotaxis is at work when newcomers arrive at a party or first enter an empty restaurant and instinctively stand at the edge for a while to take in the scene, and then select a seat at the periphery. Anecdotally, it is rare to find someone seated in the center of an empty restaurant; they will usually dine more comfortably off to one side.

Figure 2.11 Shoppers stroll in Kapalicarsi, or Grand Bazaar, an ancient market center in Istanbul, Turkey. It is essentially a covered street (Source: Wikimedia Commons, author: espiritu_protector).

Thigmotaxis in Action: Three Case Studies

In the three case studies that follow, we look at thigmotaxis in action in different development projects in the United States, and hone in on the impact continuous building alignment can have on a project's success. None of the developments in the discussion that follows evolved organically. All were at one time improbable dreams in the minds of creative, visionary, and somewhat relentless entrepreneurs who had distinctly social agendas. Making money was certainly part of their plans, but it never appears to have been the only point. Each of these businessmen seems to have wanted his new project to establish a new paradigm for building, one that would enrich the lives of college students near campus, or increase the options for middle class suburban home ownership, or invent a new way for families to entertain their children on vacation. All were financially risky ventures at one point, and designed and constructed in the early-to-mid-twentieth century. For each development, we show figure-ground drawings and photos of the final results, and then discuss how they meet, failed to meet, or exceeded anticipated outcomes. For it turns out in new developments edges matter—more than one might think.

Case Study: Palmer Square, Princeton, New Jersey

Palmer Square will fool you. Located just outside the Princeton University campus, it looks like an old town center that organically evolved with a post office, shops, and residences closely clustered together. Its development most definitely did not happen this way. Palmer Square was completely planned, the brain-child of Edgar Palmer, Princeton class of 1903, who thought the university experience would be much improved by a place for students, locals and visitors to drink, dine, and otherwise recreate, a short walk outside the school's 350 acre (142 ha) bucolic campus.

As president of the Princeton Municipal Improvement, Inc., Palmer first unveiled his scheme in 1929 (see Figure 2.12). It was to be a mixed-use project modeled after a European village, complete with a retail, office, and residential program, including a theatre and hotel. The stock market's collapse put the breaks on his dream that year, but Palmer was persistent and ground-breaking resumed several years later, in the 1930s in the midst of the Depression. In the seventy-five years since, the plan has evolved and filled in along the lines he established but did not live to see realized. Today, Palmer Square is often held out as an example of urban planning excellence in the US. The Square "over the years has blossomed into one of the finest examples of a commercial downtown," reads a typical commentary (Vilotti 2013).[3]

Palmer's architect, Thomas Stapleton, "assembled a 'potpourri' of favorite styles," to keep things visually interesting for pedestrians, explains Jerry Ford, an architect in Princeton today, describing why the square works so well. "There is a bit of old Newport, Philadelphia, Annapolis and Williamstown," in the facades (Holt 2012). The plan's continuous building alignment makes for effortless accessibility. Some 20 stores are within a 150 meter (500 feet) walk along Palmer Square West, one of its principal streets (see Figure 2.13 and 2.14).

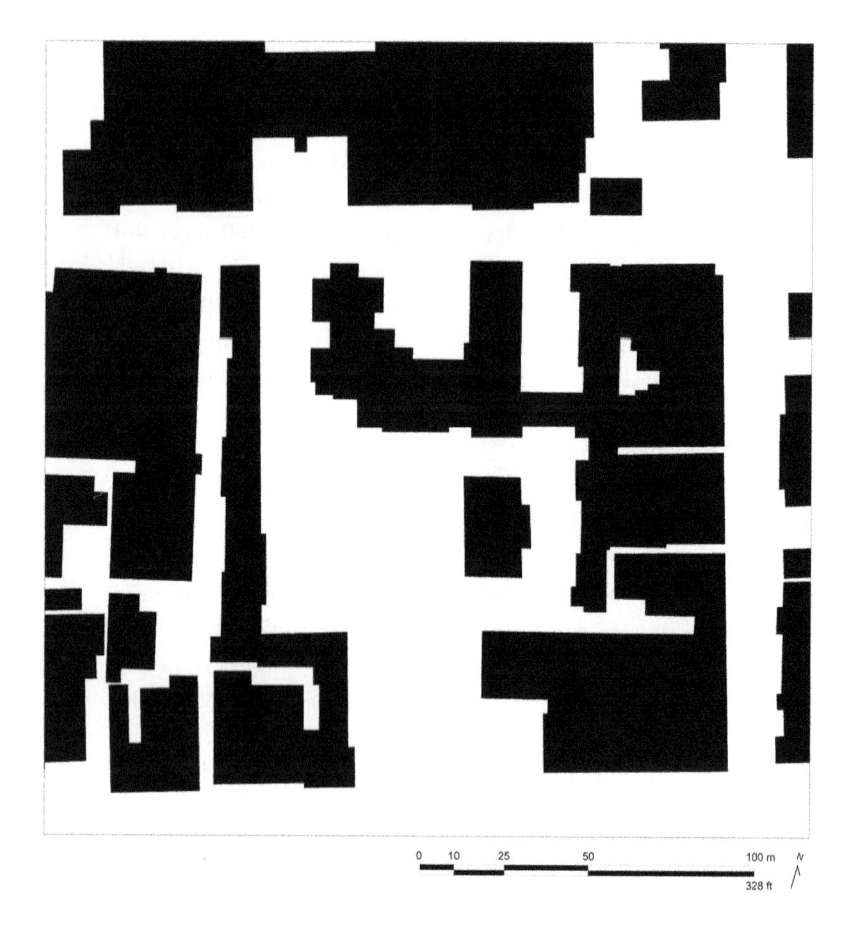

Figure 2.12 Figure-ground drawing of Palmer Square, Princeton, New Jersey, a tight, walkable shopping district under 75 meters wide that was built off a major thoroughfare (Source: Nora Shull).

The designer and developer also realized how important it was to keep busy car traffic at bay. Stapleton moved the square away from the main street, doing something not usually found in a European village, but modeled after another famous urban

Figure 2.13 Diverse storefronts aligning along Palmer Street West in Palmer Square across the street from Princeton University in New Jersey (Source: Justin Hollander).

Figure 2.14 A park between Palmer Street West and Palmer Street East is tightly enclosed by retail buildings and restaurants, making for a quiet pedestrian realm (Source: Justin Hollander).

project, Rockefeller Center, which happened to be under development in New York City at the same time, (and would become that city's largest commercial development.) "The plan of the Square," said Ford, "is a mini version of Rockefeller Center. Both were built within the decade of the thirties and both were designed to turn the commercial traffic in from a major road. In the case of Princeton that road was Nassau Street and in New York it was Fifth Avenue. The early plans for Rock Center contemplated an Opera House at the end while Palmer had the Playhouse movie theater" (ibid). In both, anticipating and accommodating pedestrian needs for well-defined edges and enclosed outdoor spaces has paid off, financially and in the positive urban experiences the centers have provided its denizens for decades.

Case Study: Columbia, Maryland

As a city-sized planned community, Columbia, Maryland seemed to have everything going for it. It was located in a prime spot in the Northeast corridor, half-way between Washington, D.C. and Baltimore, Maryland, and had James W. Rouse, a visionary real estate developer, at its helm. Rouse was an idea-person who wanted to create a new, 100,000-person city based on an altruistic premise: enhance the quality of suburban life, promote community, and combat that new scourge in the post-war American landscape: spotty development and sprawl. With his distinctly social agenda, Rouse believed tapping into social science and the latest psychological research of his era (the 1960s) could provide a clear path forward. He explained why:

> There is no dialogue between the people engaged in urban design and development and the behavioral sciences. Why not? Why not bring together a group of people who knew about people from a variety of backgrounds and experience to view the prospect of a new city and shed light on how it might be made to work best for the people who live there.
>
> (Bloom 2001: 36)

So Rouse assembled a 'Work Group' of social science professionals, including academic planners. They met regularly for over a year, analyzing the latest research on how to create the ideal city. The goals and design guidelines they came up with were intended to help future residents get along, embrace diversity including economic differences and promote social cohesion. With this Work Group, Rouse laid out Columbia's original system of nine villages, all arranged around a core 'Town Center,' which featured a man-made lake, indoor-shopping mall, and high-rise office buildings a mere 1,000 meters (1,093 yards) from a major highway on-ramp (see Figures 2.15–2.18). This put the new city, officially unveiled in 1967, in

Figures 2.15, 2.16, 2.17 and 2.18 Figure-ground diagrams of three of Columbia Maryland's ten planned villages at top and bottom right; bottom left 'Town Center' mall. All show car-dependent design, a hallmark of mid-twentieth-century design. (Source: Nora Shull).

an ideal location under an hour's drive from the nation's capital. Columbia was thus complete with jobs, schools, medical facilities, and multiple housing options including single-, multi-family, and apartment offerings, in one prime spot.

But though Rouse's group did produce 'villages' with a mix of housing types each around its own shopping zone and built the Town Center at the city's core near a lake, the urban design was in no way traditional. As the figure-ground diagrams show, Columbia's roads are curvilinear with few intersections. They favor cars. As 2D drawings and abstract designs they are interesting, even beautiful. In their 3D iteration at full scale, however, they fail completely, particularly when it comes to anticipating the needs of an upright, walking mammal (see Figures 2.19–2.22). Even though there are sidewalks in Columbia, the open landscapes around them provide little in the way of a street wall, making it difficult for people to figure out where they are and which way to head. Pedestrians and even newcomers traveling by car perennially feel lost in Columbia. The lack of street corridors and grids became the progressive city's Achilles' heel.

In 2011, a Brookings Institution land-use expert warned residents that if Columbia did not "evolve to become more walkable, it risks becoming irrelevant and declining over the next century," a local news service reported (Hazzard, 2011). Christopher Leinberger, a senior fellow at the non-profit think tank in Washington D.C., did not mince words in his talk to 450 residents, titled "21st Century Development Trends: How Will Columbia Measure Up?" "Columbia is the pinnacle of the drivable suburban option," he said. But, in the last decade the pendulum of public opinion has swung far from this model. It is likely to move even farther out in the future. Fueling the trend, Leinberger said, is the fact that the younger adult generation, the Millenials, postpone marriage and children, and enjoy the vitality of city life. Additionally, the high environmental cost of cars and the financial burden of

Figure 2.19 Columbia, Maryland's 'villages' carry few distinguishing features (Source: Nora Shull).

Figure 2.20 Acres of parking lots define Columbia, Maryland's 'Town Center' (Source: Nora Shull).

Figure 2.21 Columbia, Maryland's design makes navigating by car or on foot a challenge (Source: Nora Shull).

Figure 2.22 It can be hard to tell where you are in Columbia, Maryland (Source: Nora Shull).

maintaining a multi-car lifestyle is turning many age-groups away from the suburban option. The evidence for the shift is clear in the real estate market, Leinberger said. "The most expensive housing in the country today on a square foot basis is walkable urban. This became the case over the past 10 years and it's the first time since the 1960s."

The problems with Columbia's street layout came to a head, and have caused the most resident distress over the years, at the development's core, the Town Center. Dominated by scattered high rises, parking lots, and a large indoor mall surrounded by acres of tarmac, it is not a pedestrian-friendly place. As urban historian Nicholas Bloom reported:

> By the 1980s, in many residents' eyes, the town center as a whole had failed to achieve the level of urbanity envisioned in original publicity materials. The plan, like most modernist plans, had separated the community college, the shopping mall, office buildings, and public spaces into individual zones. Large open areas and sprawling parking lots diluted an urban feeling and made walking difficult.
>
> (Bloom 2001: 50)

It is this central district residents hope to now fix, more than forty years after it opened. They approved a new Town Center development plan three years ago, which calls for denser, mixed-use buildings to replace the acres of parking lots around the mall and make it easier for people to walk. Ironically, Rouse knew how to build walkable places, at least indoor ones, all along. He designed some of the first indoor suburban shopping malls in the U.S., realizing that residents in far-flung suburbs still needed quasi-centers for shopping, walking, meeting, and dining. He modeled the interior of Columbia's indoor Town Center mall after a traditional Main Street, lining it with diverse storefronts and even fitted it out with gas-lit lamps, wood benches, and tall old-fashioned street clocks (see Figure 2.23). It would be the only part of the planned city based on a double-loaded corridor street. So, the answer to making Columbia pedestrian-friendly was there all along. The case study illustrates the steep price residents end up paying when developers and their designers do not consider that people appreciate well-defined edge conditions indoors and out. It is a price that Columbia's residents are paying now, and it seems, per Brookings Institution studies, will be paying for some time to come.

Figure 2.23 The 'Town Center' indoor mall features Columbia, Maryland's only plan that mimics a traditional double-loaded street (Source: Nora Shull).

Case Study: Main Street, Disneyland, Anaheim, California

The only theme park designed by the cartoonist and business entrepreneur Walt Disney himself, with the help of his studio animators, Disneyland opened in Anaheim, California, on a 160 acre (65 ha) parcel in the summer of 1955. By 1965, it was listed as the top tourist attraction in the U.S. (Goldfield 2007). It has cumulatively totaled more than 650 million visits since its first day, more than any theme park in the world. Disney developed the concept spurred by disappointing visits to amusement parks with his own children, and the realization that visitors to his Burbank, California studios were eager to see more than a conventional studio tour offered. Since Disneyland's opening, imitators and the Disney company itself have spread his concept worldwide: a Disneyland opened outside Paris in 1992; another in Hong Kong in 2005. A major element of the original plan for Disneyland and contributor to its success is Main Street, USA. The twenty-meter (21 yards) long street, less than eight meters wide (nine yards), serves as the major pedestrian corridor linking the theme park's entry gate to its central focal point, Sleeping Beauty's turreted castle (see Figure 2.24). Main Street is a stylized Victorian version of the town of Marceline, Missouri where Disney grew up. The street has been critiqued as the "architecture of reassurance" (Marling 1997), a charge Disney would likely have embraced. He specifically chose the nineteenth-century style of the roadway as a historical anchor where he felt visitors, perhaps recalling a slower-paced America, would relax, and commented:

> For those of us who remember the carefree time it recreates, Main Street will bring back happy memories. For younger visitors, it is an adventure in turning back the calendar to the days of grandfather's youth.
>
> (http://en.wikipedia.org/wiki/Disneyland)

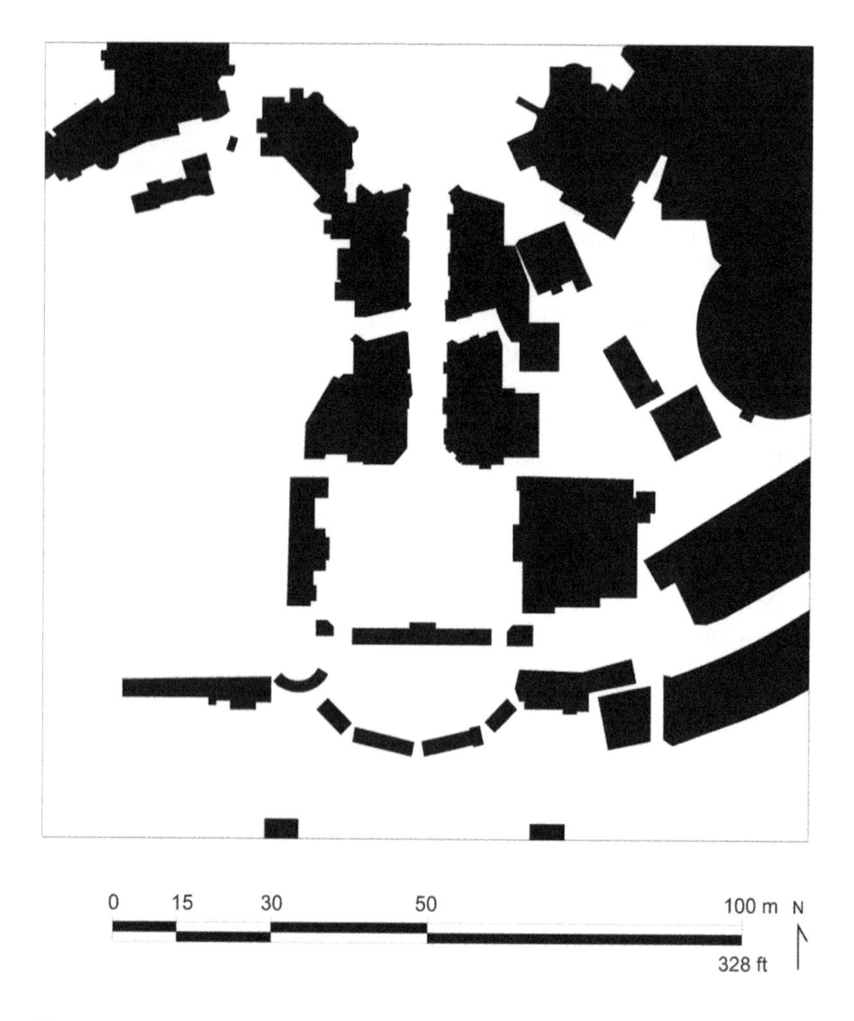

Figure 2.24 Figure-ground diagram of Main Street, Disneyland, in Anaheim, California (Source: Nora Shull).

Disneyland's Main Street is narrow, and just as Jane Jacobs noted, car-free, yet pedestrians head toward its sides. The buildings along Main Street are carefully detailed, employing a stage-design technique called forced-perspective, a strategy

Figure 2.25 Main Street, Disneyland, in 2005. Cinderella's Castle in the center distance provides a landmark; pedestrians tend to stick to the sides of the street even with few vehicles present (Source: Alfred A. Si, 'Own work,' Wikimedia Commons, July 4, 2010).

where upper floors decrease in scale, making the structures seem more imposing and appealing. The Main Street storefronts anticipate and successfully engage the visitor's gaze. It seems Disney intuited modern research, and knew that a carefully detailed row of colorful shops would not only entertain, but would help visitors situate themselves, feel secure inside the gates, and propel them on through the park in a happy frame of mind.[4] Disney, ever the animator, also seems to have known what research discussed in the next chapter has revealed, that *people love looking at faces*. (At some level, how could cartoonists not know this? Their job depends on it.) Humans are, by and large visual creatures, and as the next chapter outlines, not only are drawn to intricate details in buildings, they 'see' abstract faces in them, even assembling them from elements in the elaborate facades of storefronts Disney designed for Main Street.

At this juncture in architecture and planning history and practice, the move to replicate Disney's or similar versions of the 'Main Street' plan has moved beyond the theme park gates. Current urban planning manuals, including *Pedestrian and Transit-Oriented Design* by Ewing and Bartholomew, published in 2013, (by the Urban Land Institute and American Planning Association), emphasize the importance of re-introducing traditional Main Street plans in cities and towns, of laying out 'street-oriented buildings' and 'grid-like street networks,' of avoiding empty lots or dead zones in pedestrian centers. The U.S. Green Building Council (USGBC) has gotten into the act. Intent on promoting walkability and the development of less car-dependent neighborhoods, the USGBC's latest guidelines award projects more points in its rating system for increasing intersections and bolstering suburban and urban connectivity (the USGBC recommends 140 intersections per square mile). Urban design theory has moved on from its earlier twentieth-century predisposition for limiting intersections and pulling doors and windows off the street.

While Chapter 2 looked at thigmotaxis, our wall-hugging tendency, as a hidden driver of human navigation and movement through space, Chapter 3 explores our principal sense: vision and the significant visual pattern we arrive in the world prepared to seek. The central premise here is that human beings evolved in nature to see things that were most important for their past survival and that there is one main object for our species: the face. We have evolved to seek it subconsciously, and this built-in behavior affects not only how we behave socially but how we feel about our surroundings, behave in public places and even how we choose cars or houses and evaluate and price our art.

Exercise for Chapter 2: Thigmotaxis

- Thigmotaxis is thought to have survived for millenia because it provides animals with a way to engage with their environment. What other benefits may have promoted the trait's success across species?
- In the drawings, sketches, or photographs taken of urban or suburban areas for the Chapter 1 exercise, where can you hypothesize thigmotaxis may be at work?
- When designing a building or urban plan, what advantages do double-loaded corridors carry in general over a single-loaded plan? Why?
- Jacobs, Alexander, and other urban observers write about the significance of 'windows on the street,' meaning usually the first floor; what might be some of the reasons this is a hallmark of successful pedestrian urban environments?

Notes

1 To understand where Le Corbusier is coming from, it is useful to know that he witnessed a city whose population had more than quadrupled in a century; Paris went from 650,000 people in 1800 to 3 million by 1910. France at the turn of the twentieth century also led the world in automobile production; the city's congestion with vehicles and people seems to have been unparalleled. At the time, doing away with streets may have seemed to some a straightforward solution, at least in theory.

2 The Paleolithic time, considered to have been 2.5 million years ago to about 10,000 years ago, depending on the source, is also referred to as the Old Stone Age. Early hominids living during this prehistoric time included *Homo erectus*, *Homo neandrethalis* and emerging some 200,000 years ago, *Homo sapiens*.

3 Palmer Square is also held out as an early example of gentrification, highlighting tensions and trade-offs endemic to planning. Edgar Palmer forcibly dislocated a black community living next to the University to build the square. Their housing was destroyed, and although replaced elsewhere, it was further from the university where many residents worked, making their commutes more difficult. The project could be viewed as pitting one class's interests against another's, an aspect of the development apparently less discussed.

4 The official slogan for Disneyland is 'The Happiest Place on Earth.'

3

Patterns Matter: Faces and Spaces

> Human history can be viewed as a slowly dawning awareness that we are members of a larger group.
>
> Carl Sagan

Faces. We do not look out at the world as though all things in it are of equal value. We look through an evolutionary scrim that prioritizes faces and people. We are a social species, after all, which spent more than 99% of our time on earth in the wild and not in man-made environments (Kellert 2012, x). The importance of perceiving and connecting to faces in our lives and in our evolution, whether those of humans or other animals, is hard to overstate (see Figure 3.1). The theory goes that if we had not evolved to prioritize faces to this extent, we would not have survived until now—at least not in our current form.

This chapter discusses the science around human face-sensing abilities. It argues that modern research in face-perception is significant for the humanities and, particularly, architecture, planning, and aesthetics. It suggests these findings provide a new type of foundation for the building-disciplines with implications for understanding why people favor certain buildings over others or even why they head to certain streets and city squares and avoid others. To design successful places for people it is important to have a basic understanding of how people work, and one of the prime things people do, it turns out, is look at other people.[1]

Figure 3.1 One-year-old Thomas connecting to his grandfather Martin; humans have evolved to prioritize vision and, within that same category, the face (Source: Ann Sussman).

The Senses are Not Equal

To appreciate the extent we prioritize seeing faces it is critical to understand that the human senses are not equal or they do not carry equal weight in our perceptual apparatus. We have five basic senses: smell, hearing, touch, taste, and vision, and the human brain expressly prioritizes one of them: vision. Our brain works hardest at creating our visual view of our surroundings, explains Eric Kandel, the neuroscientist and Nobel prize-winner in *The Age of Insight: The Quest to Understand the Unconscious in Art, Mind and Brain from Vienna 1900 to the Present* (2012). "We are intensely visual creatures and live in a world oriented to sight," he wrote, which he goes on to say, can explain why paintings can have such an emotional impact on viewers. "Half of the sensory information going to the brain is visual" (Kandel: 238). We are very good at seeing and interpreting faces, and portraits wordlessly acknowledge that talent. In a way, we behave like kids in a classroom, always ready to do what we are good at.

The Brain's 'Rules': We See What Our Brain Wants Us To See

A major point Kandel makes is that the brain "reconstructs reality according to its own biological rules." (Kandel 2012: 301). While the 'eye is a camera,' may be a popular way of explaining how light, reflected off objects enters the eye where it stimulates the light-sensing nerve cells (rods and cones) at the back of the eye (the retina), this camera model does not account for the fact the processing of what we see and feel happens in our 'gray matter,' or cerebral cortex—not the eye. The brain particularly appreciates portraits because they show faces that it has evolved highly specialized circuitry to interpret. "Neuroscientists found that one reason the face is so important for perception is that *the human brain devotes more area to face recognition than to the recognition of any other visual object*" (Kandel 2012: 333) (italics ours). That sentence bears repeating. We are face readers, par excellence. Nature determined long ago what pattern would be most important in our visual view of the world, the one that, not coincidentally, was also most important for survival.

Knowing this bias, neuroscientists note, with some amazement, how each of our individual mentally synthesized versions of a face, and of reality generally, is able to mesh with someone else's. "Each of us is able to create a rich, meaningful image of the external world that is remarkably similar to the image seen by others." Kandel wrote (2012: 234). Our brains create "a fantasy that coincides with reality," the English psychologist Chris Firth elaborates, explaining that inputs from our senses to the brain produce something "much richer" than what is actually there: "a picture that combines all these crude signals with a wealth of past experience." (Kandel 2012: 234) The fact we enjoy portraits hints at the tremendous amount of work the human brain does outside of our awareness. The fact that art engages us—and not your pet cat, for instance—speaks volumes about the human brain's uniqueness. Artists are able to emotionally involve the human viewer and enable him or her to empathize with their work by incorporating and stimulating the 'beholder's' mind, Kandel writes. Somehow artists know how to get to the pieces of our brain we do not see or even realize we have, for that matter, and then manipulate them, arguably for mutual benefit. Thus, a painting exists in the moment when viewed, creating an impression that occurs as an effect of cognitive processing, all occurring without the viewer's conscious control. From this perspective, visual art exists as a tell-tale artifact broadcasting our mind's mysterious functioning (see Figure 3.2).

The Brain's 'Rules': A Template for the Face

In the past twenty-five years researchers have started to untangle some of the complex processing that occurs in these secret reaches. Until relatively recently, whether facial processing itself was learned or innate was debated. From the 1970s to the 1990s a dominant view in the scientific community was that children required years of experience to attain adult-like facial-recognition capabilities (McKone *et al.* 2012: 2). However, the consensus now has emerged that we arrive in the world more or less ready to process faces, or "experience is less important than previously believed," as McKone *et al.*'s 2012 article in *Cognitive Neuropsychology* noted. "The evidence clearly indicates that the

Figure 3.2 Self-portrait, Grimacing, 1910, by Austrian painter Egon Schiele (1890–1918); Austrian artists like Schiele worked at revealing the inner states of their subjects, argues neuroscientist Eric Kandel. The paintings engage the viewer subconsciously and our involvement can be understood as an artifact of our brain's unique architecture (Source: Wikimedia Commons).

ability to encode faces is present very early in life," particularly, when presented right side up, writes scientist Elinor McKone and her colleagues. "Babies discriminate faces upright but fail to discriminate the same stimuli inverted" (p. 7). Later in this chapter, we look at this inversion effect as a way to appreciate some of the brain's 'rules' of perception. Nature has preset not only the principal pattern for our visual field but also set its expected orientation. Evolution brings to the present what worked out in the past.

The advent of functional magnetic resonance imaging (fMRI) in the 1990s proved a game changer in terms of uncovering more of the brain's tactics for face-processing. By charting the change in magnetic fields, fMRI tracks blood flow to specific brain areas. Since neurons require more energy when activated, delivered by the blood supply, the increased blood flow can show what happens in the brain as people view different things, such as when they look at faces as opposed to other objects.

Results from early fMRI brain scans led researchers to discover a specialized module for faces (in the temporal lobe of the cerebral cortex) that became known as the fusiform face area (FFA). Scientists have since theorized this dedicated processing place may be present in all mammals, enabling speed and more: "it makes evolutionary sense to have a face system capable of rapid, accurate face recognition from an early age to support social development, and ultimately survival" (McKone *et al.* 2012: 31). An early and current leader in face-perception research includes neuroscientist Nancy Kanwisher, now at the McGovern Institute for Brain Research at the Massachusetts Institute of Technology (MIT). In a 1997 experiment, she gave twenty test subjects fMRIs while they quickly looked at faces and other objects including scrambled visages. She found that only one specific region in the brain's temporal lobe "produced a significantly higher signal intensity during epochs in which faces were presented than during epochs in which objects were presented" (Kanwisher *et al.* 1997: 4304). This region, the FFA, "responds more strongly to faces than objects" (Kanwisher *et al.* 1997: 4308). Further, the existence of the FFA suggests "qualitatively different kinds of computations" occur in facial processing than for other objects in our visual field, she wrote. In other words, faces are so significant, the brain evolved a specialized program and place for dealing with them. Kandel calls this the brain's 'template-matching' approach. Since faces, and to a lesser extent hands and bodies, can give us so much information quickly—they tell us about the age, health, sex, attitude of an individual, at a glance—the brain does not put facial or body inputs from "a pattern of lines," or the part-based processing that it uses with other visual inputs (Kandel 2012: 287). Because part-based processing is relatively slow, face-processing evolved using a different technique, a pre-existing pattern. Kandel labels this the "figural primitive," and describes it as an oval, right-side up, with two points for the eyes, a vertical line for a nose, and a horizontal line below for the mouth (see Figure 3.3).

Figure 3.3 'Figural Primitive' of the face used in human visual processing from infancy on as rendered by artist Trey Kirk (Source: Trey Kirk).

Recent fMRI research has also found the location of other regions of the brain believed to be preset for viewing the body[2] and a specialized area for viewing natural landscapes or interiors.[3] There appears to be more evolutionary logic at work here. "It is crucial to recognize and interpret information conveyed by the bodies of other members of one's species," according to *Cognitive Neuroscience*, a current college text (Banich and Compton 2010: 201). "This involves recognizing not just their facial identities and expressions, but also their bodily movements, postures, stances and gestures." The fact the brain comes pre-equipped to respond to several classes of objects has led some neuroscientists to describe it as a biological 'Swiss-Army knife,' ready to efficiently process specific inputs once these present themselves in our visual field, much as a jackknife's corkscrew anticipates a cork or its screwdriver is ready to fit a screw (Tech day, 2013).

The Brain's 'Rules': Faces and Bodies Right-side Up

Researchers also note how human body-processing by the brain appears to resemble face-processing in its sensitivity to orientation. "(I)nversion disrupts the recognition of bodies, just as it does for faces; in fact, the effects of inversion are similar for body postures as for faces, with both showing a larger effect than nonbiological categories" (Banich and Compton 2010). One way to observe the brain's bias for viewing faces right-side up easily is to look at the work of the sixteenth-century painter Guiseppe Arcimboldo (1527–1593). A painter to the Viennese court, Arcimboldo worked in a creative 'Mannerist' style, a late-Renaissance approach known for creatively exploring man's place in nature.

The paintings on the following page (see Figures 3.4 and 3.5) appear to be a bowl of fruit or some sort of fruit pile. Flipped

Figure 3.4 The Gardener, *c.* 1590, viewed upside down, by Italian painter, Guiseppe Arcimboldo (1527–1593). A Mannerist, Arcimboldo worked in a transitional style between High Renaissance and Baroque, one known for intellectual playfulness (Source: Wikimedia commons).

Figure 3.5 The Summer, *c.* 1593, upside down, by Arcimboldo (Source: Wikimedia Commons).

180 degrees, or right-side up, displayed the way they were intended to be seen, the images snap into place as a human face and profile (see Figures 3.6 and 3.7) (These oil portraits were also apparently likenesses of real people, illustrating Arcimboldo's considerable skill).

Another example of how our brain wiring prioritizes faces right-side up is the 'Thatcher Effect' or 'Thatcher illusion'. In 1980, Peter Thompson, a psychologist at the University of York, observed that it was difficult for people to interpret expressions in an upside-down face and that other researchers had noted similar findings. So Thompson obtained a photograph of Prime Minister Margaret Thatcher, the leading and controversial UK politician of the decade, and turned her features right-side up and then placed these on her upside-down photo (see Figure 3.8).

Figure 3.6 The Gardener, right-side up (Source: Wikimedia Commons).

Figure 3.7 The Summer, right-side up (Source: Wikimedia Commons)

Would her upside down face be easier to read with her features right side-up? he wondered. No, he found out. It did not much matter whether the lips were right-side up or upside-down on the inverted face: "such transformation makes little difference to Mrs. Thatcher's expression (Thompson 1980: 483)." Turned right-side up, however, the viewer is in for a shock. With its upside-down eyes and mouth on her upright photo, she became unexpectedly "grotesque" and strangely riveting (Dahl *et al.* 2010). It appears "we have been cruelly deceived by the smiling Mrs. Thatcher," Thompson noted wryly (ibid: 483). Though the photographs in the left and the right columns are indeed identical, we do not 'see' them that way. Our wiring is not set up to process the inverted image the way it is set up to process the upright one. We have been fooled; nature had us again.

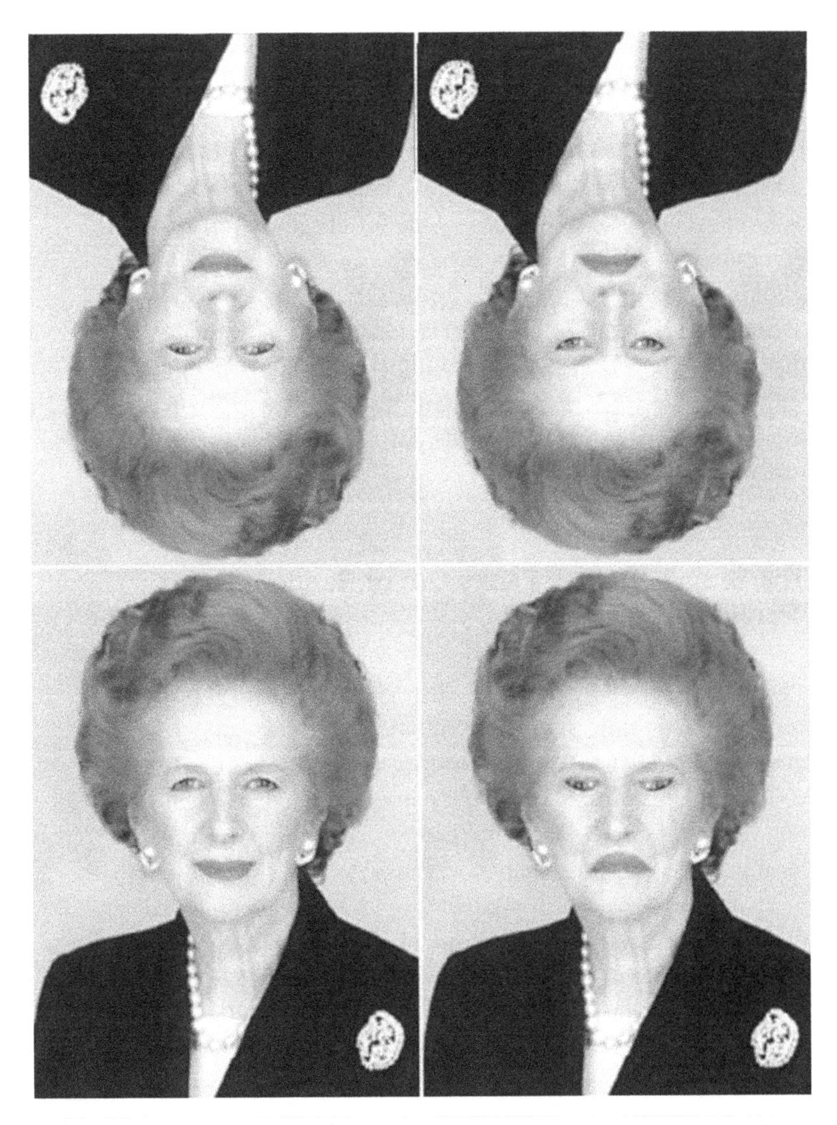

Figure 3.8 'Thatcherized' images of Prime Minister Margaret Thatcher by artist Nora Shull from an official photograph. Human subjects will focus on the face on the lower-right-hand corner, even though the one directly above it, in the top right, is identical. Our processing mechanism prioritizes the right-side-up face.

Over the next thirty years, Thompson's one page article in *Perception*, would become one of the most referenced articles in the English psychology journal. It would influence research in related species. Scientists studying rhesus macaques found that these primates "rely on the same mechanism of face perception as humans do, i.e. holistic processing for upright faces." Dahl *et al.* wrote in a 2010 article in the *Proceedings of the Royal Society* (Biological Sciences). The investigators presented monkeys with 'thatcherized' monkey faces, tracked their eye movements and discovered the primates spent more time looking at monkeys with 'thatcherized' features than those without them; the monkey responses, in other words, parallel our own, as might be expected of a related species.

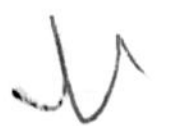

The Brain's 'Rules': Faces Out of Random Data, 'Pareidolia'

The human brain's adeptness at processing faces appears to contribute to quirks in our perception where we easily see faces in places they are not. We find faces in clouds, the moon, a tortilla chip, or the burnt markings on a piece of toast. This phenomenon is called pareidolia and comes from the Greek words meaning 'wrong' and 'shape'. It refers to how we effortlessly make illusions from random data. A well-known pareidolia example from outer space is at right, the photograph taken by NASA's Viking 1 Orbiter as it flew over Mars in 1976 (see Figure 3.9).

Was this the work of Martians? The image caused such buzz NASA sent a follow-up mission to the site on Mars a decade later to take higher resolution, close-up photos and—spoiler alert—reveal this Martian 'mesa' or hill had no human or actual facial attributes whatsoever. Yet outside of fueling conspiracy theories, biologists consider pareidolia evolutionarily advantageous. It is better to be able to recognize a face in poor visual conditions than to miss out on one entirely. If you do not recognize the

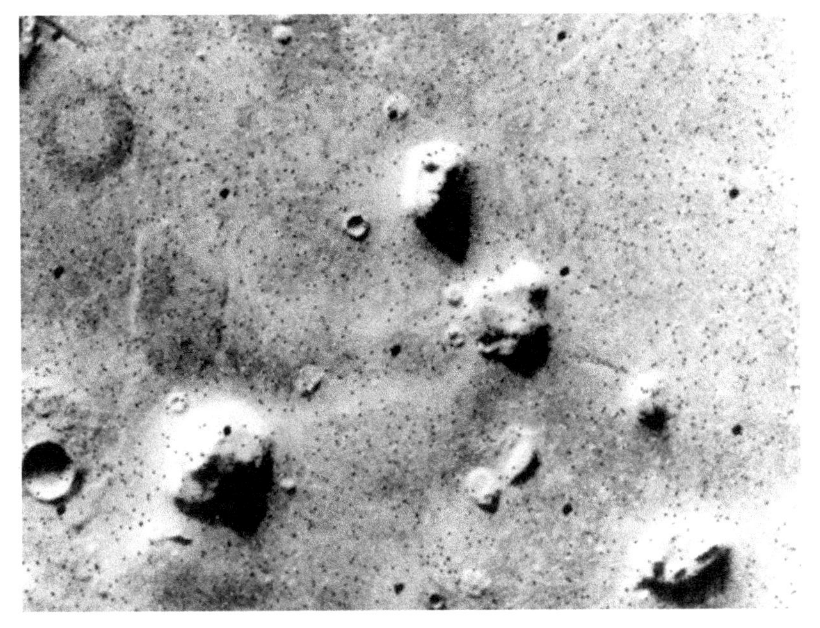

Figure 3.9 The photograph of a Martian hill or mesa, taken from NASA's Viking 1 Orbiter as it flew over Mars in 1976, seems to show a human face (Source: Viking 1, NASA; Wikimedia Commons, December 27, 2010).

snake in the grass it may prove fatal; mistake a blade of grass for a snake and much less harm done. The astronomer and popular science writer Carl Sagan hypothesized why humans may have such propensity for pareidolia in his book, *The Demon-Haunted World: Science as a Candle in the Dark* (1995):

> As soon as the infant can see, it recognizes faces, and we now know that this skill is hardwired in our brains. Those infants who a million years ago were unable to recognize a face smiled back less, were less likely to win the hearts of their parents, and less likely to prosper. These days, nearly every infant is quick to identify a human face, and to respond with a goony grin.
>
> (Sagan 1995: 45)

Infants over generations have been 'selected' for facial-recognition to such an extent that today members of our species are not only face-experts, they see visages everywhere, well past their early childhood and throughout adulthood. Natural selection can be seen at work.

The extent to which our facial-processing templates are also used to process face-like non-living objects is a matter of on-going research. According to one recent study the same FFA that permits us to quickly process our mom's or our mate's face may be used in processing inanimate objects, such as car grills (Banich and Compton 2010: 200) Researchers studying car experts, for example, found that when the auto afficianados were briefly presented (for three-hundredths of a second) with images of car fronts and profiles the same FFA responded as for human faces. How generalized this effect is for other objects is open to debate and further study.

The Brain's 'Rules': Faces Engage Emotions and Memory

What is clear is that even when the object is not human, a face-like object engages us emotionally without conscious effort. We distinguish an animate face from an object's face, yet still have feelings toward it. By design we remember faces, although not necessarily names, and it appears this trend spills over to face-like objects. The tractor on the following page (see Figure 3.10), for instance, seems to have an alert wide-eyed youthful gaze; one might find it attractive and lovable, even. The machine's face is memorable. The tractor's owners, no surprise really, gave it a name, and treat it as a family member.

Car manufacturers know people form emotional attachments to automobile 'faces' and use the fact to guide their design and marketing efforts. "In today's hyper-competitive car market, designers are focusing on faces as part of a broader effort to

Figure 3.10 'Robert the tractor,' a 1957 Ford 661 Workmaster (Source: Ann Sussman) is a beloved member of the family for owners in Duchess County, NY.

design cars that appeal to buyers—tapping psychologists, anthropologists and other experts in human behavior, and even monitoring the brain waves of focus-group participants," the *Wall Street Journal* reported in a 2006 article entitled, 'Why Cars Got Angry.' Automotive research shows "70% of drivers identify and judge vehicles by the headlights and grille," the article reported. In the mid-2000s the trend was toward angrier, scarier vehicles visages. Perhaps, the article hypothesized, to help drivers feel safer in heavier traffic amidst more oversize SUVs.

A different but also familiar example of how faces quickly connect to emotions can be found in email messages that use emoticons, the typographical notations shown below:

:) happy
:-(sad
;-) wink

When a writer adds the marks to an email text, we instantly have a clear idea how she or he feels about the subject. The fragmentary lines appear to us as facial features and work like emotional shorthand. Scott Fahlman, the Carnegie Melon computer science professor credited with inventing emoticons in 1982, did so to help readers distinguish serious posts versus silly ones on the then new computer message boards. (Before emoticons, he explained, people had no way of quickly knowing whether a post was written seriously or as a joke, resulting in many unfortunate and unnecessary misunderstandings.)

Faces are Everywhere in Art and Marketing

Knowing our brain is designed to prioritize faces, perceives them from the earliest age, and engages our emotions as it does so, makes it easier to understand why creative directors of all kinds use faces everywhere. Faces abound in art. The most famous painting in the world, and arguably most valued, Leonardo da Vinci's Mona Lisa, is appraised at more than $750 million and depicts an attractive, mysterious face. Faces also lead the evolving list of the most expensive paintings sold worldwide. Nine of the ten most expensive paintings ever sold portray faces, with Cezanne's The Card Players topping the list at $267 million and Van Gogh's Portrait of Joseph Roulain at $110 million, rounding it out. Faces dominate in advertising and retail. Below is an Apple store outside Boston in 2013 (see Figure 3.11).

Every single new iPhone featured in the store posters shows a large image of one or more faces. Retailers know at some level that subconsciously we cannot look away: they take advantage of our facial-processing predisposition. So, too, we find the face on billboards, in movies, on YouTube, and TV. Director Steven Spielberg gained fame for slow pans and close-ups of faces, a steadfast focus that helped make his name and secure his fame as a director.

Figure 3.11 The interior of an Apple store in metropolitan Boston promotes products by showing different large faces on each device (2013). Subconsciously we cannot look away. The display feeds our neural circuitry precisely, metaphorically fitting our face-orientation like a glove (Source: Ann Sussman).

Faces are in Buildings

What about buildings? Not surprisingly, a building designed to look like a face grabs our attention and can hook our emotions much like a billboard papered with one does. The Lampoon Castle, opened in 1909, just outside of Harvard Yard in Cambridge, Massachusetts. The brick building is the home of *The Harvard Lampoon*, a satirical college-student newspaper. Designed by architect and Harvard graduate, Edmund M. Wheelright, one of the newspaper's founding members, the front of the building looks like a wide-eyed youth sporting a Prussian helmet on his head (see Figure 3.12 and 3.13).

Walking toward it today, the Castle grabs your attention more than anything else in the vicinity, particularly when viewed from

Figure 3.12 The Lampoon Castle, home of a satirical Harvard student newspaper has caught the eye of pedestrians walking down Mount Auburn Street in Harvard Square, Cambridge, Massachusetts since its 1909 opening (Source: Ann Sussman).

Figure 3.13 The side elevation of The Lampoon Castle, designed by architect Edmund Wheelwright, reads like a face, too (Source: Wikimedia Commons).

the front. An early critic commented the structure, was "laughing at every turn with freakish gayety and beauty" (Wald 1983: 40). For people we have talked to, it still makes them smile. In 1978, the Castle was added to the National Register of Historic Places.

A similar approach, designed recently, is taken in The Portrait Building, in Melbourne, Australia, by architectural firm Ashton Raggatt McDougall (also known as ARM Architecture, Melbourne) (see Figure 3.14).

Scheduled to open in 2014, The Portrait Building carves out the facial features of William Barak, a nineteenth-century aboriginal leader, in the concrete balconies of a thirty-two-story, 100 meter-high, residential tower. The design pays homage to the first Australians and aims to figuratively, as well as perhaps

Figure 3.14 The Portrait Building, Melbourne, Australia, by architectural firm Ashton Raggatt McDougall, scheduled to open in 2014, memorializes the face of aboriginal leader William Barak (1824–1903). (Source: ARM Architecture).

literally, cement the link between the modern city and its original inhabitants. Like the Lampoon Castle, The Portrait Building intends to get your attention and succeeds. It works particularly well viewed from a national war memorial about 2 miles down the road. Like the billboard it essentially is, The Portrait Building speaks most clearly at a distance, its message disappearing at close range or oblique angles.

Facial Expressions in Buildings

Our face-sensing capability is so strong and present that faces also appear to be put into building elevations or facades unintentionally. One might call that tendency 'face-a-tecture.' It reflects the fact some researchers believe pareidolia, the subconscious tendency to assemble faces in random objects, plays a much more significant role in design, aesthetics, and our appreciation of buildings and cityscapes than is generally realized. Consider the vernacular buildings on the following pages (Figures 3.15 and 3.16.)

These elevations resemble Kandel's 'figural primitive' discussed above (see Figure 3.3). Bilaterally symmetrical windows with shutters look like eyes, just as in the template, while the doors, centrally placed, appear to be a nose and/or mouth. It is almost as though the designers of these buildings—and there are many structures like this around the world—were copying the pattern they knew best, the one pre-programmed into our brains and so significant for survival: the face.

We know from anecdotal experience that streets lined with face-like elevations appeal to people and seem preferred over streetscapes that lack them. Movie scouts chose Lacock Village in Wiltshire, England (see Figure 3.17) with its corridors of friendly-looking houses as a film location multiple times, twice as a backdrop in Harry Potter films, and two more times as settings for BBC television series. Lacock Village dates from the

Figure 3.15 The Dunker Church, *c.* 1852, Sharpsburg, Maryland (Source: Ann Sussman) seems to be ever observant and even mournful; it stands on Civil War battlefield of Antietam.

Figure 3.16 Bavarian Inn, a tourist stop in Shepherdstown, West Virginia, looks friendly (Source: Ann Sussman).

Figure 3.17 A street in Lacock Village, Wiltshire, England, owned by the UK's National Trust, presents a row of face-like fronts and appeals to tourists (Source: Celia Kent).

Middle Ages, is owned by England's National Trust, which preserves historic sites, and remains one of the country's most visited tourist destinations.

Not only film scouts and travelers are drawn to its streets: some computer scientists see its popularity as significant for their field. The way people subconsciously create faces from random inputs may need to be programmed into robots, they argue, to make the machines more closely approximate our responses in our surroundings, and then, in theory at least, predict our behavior and become better helpmates.

Our minds detect diverse facial expressions in buildings all the time, report Stephan K. Chalup and Kenny Hong, computer science researchers, and Michael J. Oswald, architecture professor, all of the University of Newcastle, Australia. These visual inputs likely contribute to making places appealing to people, they write. In a 2010 paper, 'Simulating Pareidolia of Faces for Architectural Image Analysis' the researchers discussed how to use computer analysis to track the possibility of perceiving a wide range of feelings in architectural elevations including "sad, angry, surprised, fearful, disgusted, contemptuous, happy, (and) neutral" (Chalup *et al.* 2010: 268). Their computer program demonstrated how multiple emotional interpretations were possible in different buildings and sometimes within the same facade, depending on the way the windows, doors, and other building elements were arranged to make the face (see Figures 3.18 and 3.19). Since "the perception of faces is qualitatively different from the perception of other patterns," they explain, it can contribute to understanding why "(f)aces frequently occur as ornaments or adornments in the history of architecture in different cultures," (Chalup *et al.* 2010: 262). They conclude:

> Faces, in contrast to non-faces, can be perceived non-consciously and without attention. These findings support our hypothesis that the perception of faces or face-like patterns, may be more critical than previously thought for how humans

> perceive the aesthetics of the environment and the architecture of house facades of the buildings they are surrounded by in their day-to-day lives.
>
> (Chalup *et al.* 2010: 273)

If significant for computer science and robot development, these findings are important for architecture and planning. They expand our understanding of how existing or proposed built-environments are or will be perceived. Irrespective of architectural style, epoch, or culture, humans have evolved to be face-sensors. Our proclivity to seek, find, and remember faces, both real and imagined, is not going away. Indeed, our face-processing bent has a strange way of reasserting itself wherever we are, no matter the place or project, nor how new or note-worthy a building's designer or initial intent.

Figure 3.20 is artist Jeff Koons' 13 meter high (43 foot) floral sculpture Puppy, which dominates the entrance to the new Guggenheim Museum in Bilbao, Spain, designed by the Pritzker-prize winning California architect, Frank Gehry. The puppy's enormous face is more than 4 meters high (14 feet) and became part of the museum's permanent installation, outside its front door in 1997, the same year the building opened. The sculpture adds critical detail and visual interest to the museum's otherwise abstract and streamlined facade. Because Puppy fits our brain's face-processing predisposition like a hand to a glove, the sculpture makes the Bilbao Guggenheim much more accessible than it would be otherwise. It draws its human visitors in subconsciously. They simply cannot look away. Arguably, any other sculpture or poster of a large face in its place would do the same, particularly a baby face, which we have evolved to adore. Abstract shapes may be evocative and appealing—the preference of humans for curves over sharp, jagged shapes is discussed in Chapter 4—but they can never take precedence over the first primal pattern.

Figure 3.18 and Figure 3.19 Building faces can have diverse dispositions, sometimes within the same facade, which influence us subconsciously; the top image is from the city of Newcastle, Australia, and the bottom is from Ruit, near Stuttgart, Germany (Source: Chalup *et al.* 2010).

Figure 3.20 The appealing face of 'Puppy', by American artist Jeff Koons, both dominates and draws visitors to the front door of the Guggenheim Museum in Bilbao, Spain, by architect Frank Gehry (1997) (Source: Wikimedia Commons).

Faces and Spaces

For urban design and planning, the significance of face-processing in our mental apparatus has much broader implications. It suggests one reason dimensions in urban planning—for streets, parks, city squares, building setbacks, and the like—carry emotional weight and how scale and size come into play in determining a project's public response. Dimensions are numbers, obviously, but our perceptual apparatus does not experience them neutrally. Because people are so important to us, the extent to which we are able to recognize another person's face in a place, and a human body more generally, may act as a *de facto* marker for the space's impact. This theory is elegantly explained by Danish architect and urban designer Jan Gehl in *Cities for People* (2010). "Man is man's greatest joy," Gehl writes, quoting an ancient Icelandic poet and underscoring what Gehl feels should be a clear tenet in urban planning: people delight most in seeing other people (Gehl 2010: 23). Designers and planners who accommodate this client predisposition will make cities, towns, and public places much better for people. The dimensional parameters for human vision that determine how well we can see each other are outlined below.

Main Visual Thresholds for Reading the Human Body and Face

Depending on light conditions, our eyes can distinguish another human from a background object or animal at about 300 to 500 meters (330 to 550 yards). But this distance is too far to be considered the "social field of vision," per Gehl's research:

> Only when the distance has been reduced to about 100 meters (110 yards) can we see movement and body language in broad outline. Gender and age can be identified as the pedestrian approaches, and we usually recognize the person at somewhere between 50 and 70 meters (55 and 75 yards)... At a distance of about 22 to 25 meters (24–27 yards) we can accurately read facial expression and dominant emotions.
>
> (Gehl 2010: 34)

Keeping in mind that 'man is man's greatest joy,' the 100 meter mark, where we can make out the movement of another human, turns out to be crucial; it underlies athletic fields, stadiums, and the plans of some of the world's most visited monuments and urban sites. Figure 3.21 a/b illustrates a sports field, typically about 100 meters long, which also is the maximum viewing distance, approximately, from its stadium seats.

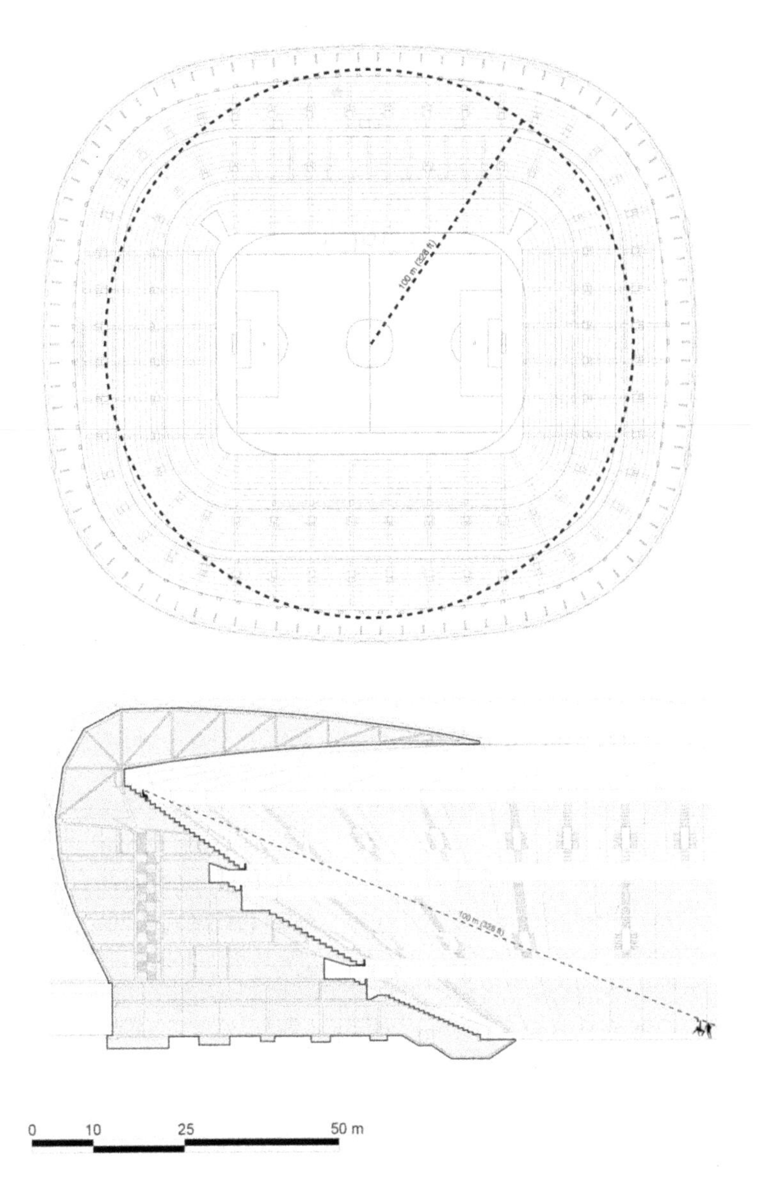

Figure 3.21 a/b Plan and section of Allianz Arena, Munich, Germany, 2005, by Herzog and de Meuron, show the 100-meter-threshold at work; the dimension sets the limit for viewer and player participation alike establishing the distance of the furthest seats and approximate length of the sports field (Source: Trey Kirk).

One-hundred Meters: The Limit of the 'Social Field of Vision'

Popular monuments across cultures use the 100 meter dimension. In the examples featured in Figure 3.22, it shows up in the Taj Mahal Gardens in Agra, India; St Peter's Square in Rome; and Places des Vosges in Paris. The 100 meter distance marks the radius length from the plan's central point as in the sports stadium in Figure 3.21.

More intimate civic spaces have 100 meters as a scale maximum, more or less. Figure 3.23 is a diagram of Piazza del Campo, the popular medieval square and tourist attraction in Siena, Italy. Note, the rectangle drawn inside the diagram for scale measures 100 meters by 60 meters.

It is interesting to compare this medieval Italian square with a modern American iteration, Boston City Hall Plaza, *c.* 1968, discussed earlier in Chapter 2. Both civic places are shown on the following page as figure-ground diagrams showing the same 100 × 60 meter (328 × 197 foot) rectangle in each for scale. (The Boston City Hall building is represented by the large black rectangle with a white square inside in the bottom drawing.)

At a glance, we can see that Boston City Hall Plaza is far outside the parameters of our social field of vision, which Piazza del Campo, on the other hand, wraps tightly around and embraces. The Boston plaza suffers the consequences of its large amorphous shape; people avoid it, as noted in Chapter 2. The city residents suffer, too; they have a civic Plaza in name only, which the *Boston Globe* described as a 'windswept urban wasteland' in a 2013 editorial. The lesson is clear: it becomes very costly for residents to fix an urban plan when it was originally laid out without regard to what people like to see (other people) and their built-in visual limits.

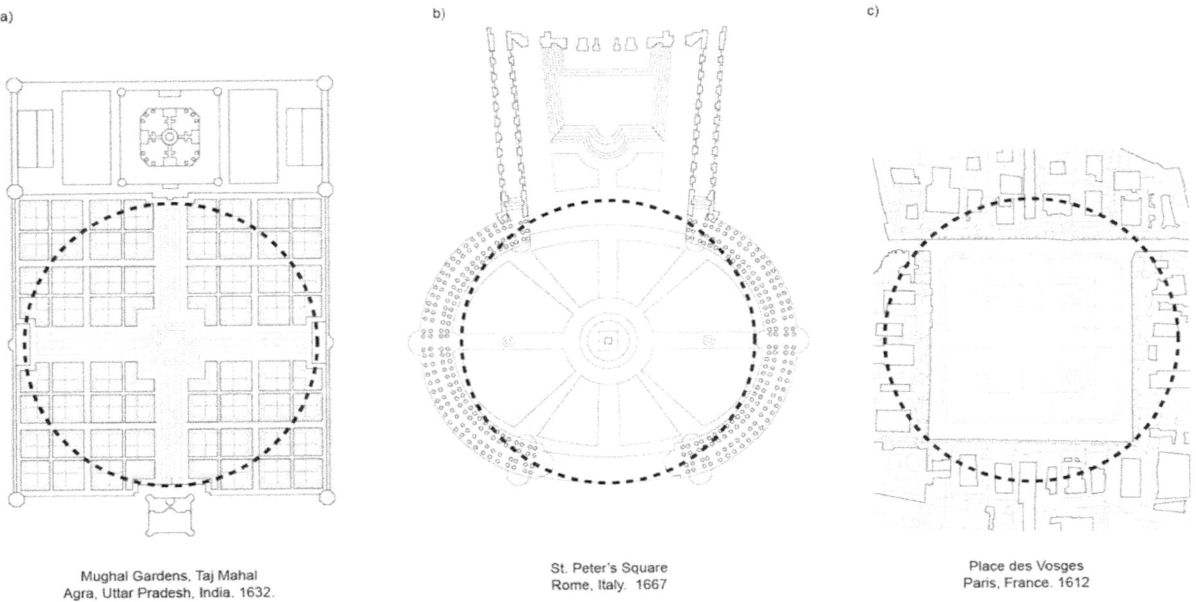

Figure 3.22 The 100-meter-threshold is embedded in the plans of many of the world's most famous civic and religious places. From the left, The Taj Mahal Gardens in Agra, India; St Peter's Square in Rome, Italy; and Places des Vosges in Paris, France. One hundred meters marks the radius length from the plan's central point to its edge. (Source: Trey Kirk).

Figure 3.23 Piazza del Campo, the popular medieval square in Siena, Italy makes for a great civic and social space. A resident walking into its center is likely to easily find and recognize a friend or acquaintance in the same space. (The rectangle delineated inside the piazza for scale is 100 meters by 60 meters.) (Source: Trey Kirk).

Figure 3.24 and Figure 3.25 Compare and contrast how the limits of our social field of vision mesh with the Piazza del Campo (Siena) on the left, but not with Boston's City Hall Plaza, on the right. The ability to see faces helps define spaces and is a consequence of the primacy of face-processing in the human animal (Source: Trey Kirk).

Thirty-five Meters: The Threshold for Reading Emotion

Another important threshold is 35 metres (38 yards) or the outer limit for reading facial expressions and emotion (without screen amplifications). The dimension comes up in theatre design—particularly in seating arrangements (see Figure 3.26). And, it turns out to be tremendously significant for laying out city streets—which, as Jacobs noted, are perennial spots for unscripted performance.

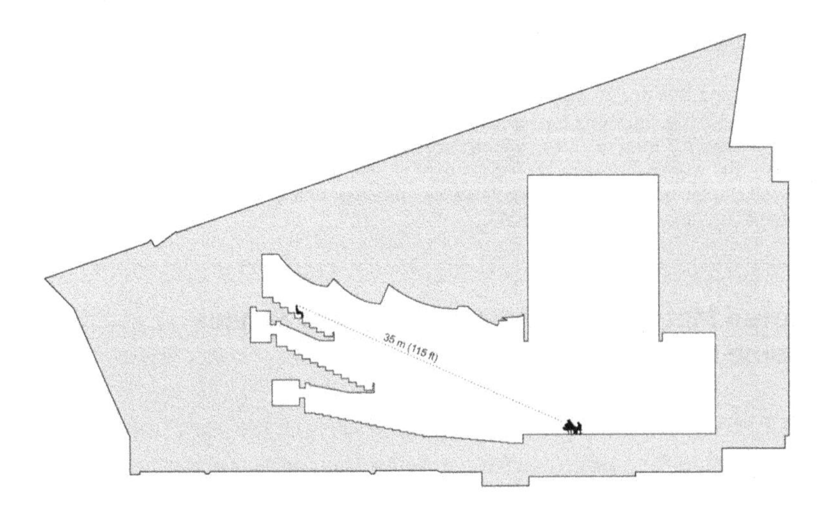

Figure 3.26 Section of the Grand Canal Theatre (2006) in Dublin, Ireland. Thirty-five meters is considered a maximum threshold for reading facial expression without electronic amplification (Source: Trey Kirk).

Facial expressions get more distinct and richer closer in. We can take in more information from 22–25 meters and under, a threshold sometimes referred to as the 'emotional field of vision.' As Figure 3.27 shows, the closer you go the more dramatically interesting things get because you see—and your mind is more preoccupied in processing—more of the face.

Figure 3.27 We find things become increasingly interesting when we can use all of our senses; this happens for humans at close range of about 7 meters (7.5 yards) or less. At about 3 meters or less we can engage in personal conversation. In the photograph, the woman's image on the far right is where our eyes will linger; the figure at far-left center, whose form is barely visible, represents the 100 meter limit, and is the least interesting (Source: Trey Kirk).

Seven Meters and Under: More of the Senses Come into Play

At 7 meters (7.5 yards) and under, things become the most interesting when it comes to viewing another human, as we are more easily able to hear someone and use all of our senses. At 3 meters (3.2 yards) and less we can engage in conversation and the visual processing is most rich. Many successful streets, the ones where people instinctively gather, are well within this 'emotional field of vision' of 25 meters and under. "When in doubt, leave the meter out," Gehl advises designers, reminding them how people cannot naturally exchange information and visually assess each other—which is what they most enjoy doing, per the ancient poet—if they are too far apart. A good example of a street that fits within our emotional field of vision is Hanover Street in Boston's North End (see Figure 3.28 and Figure 3.29), discussed in Chapter 2.

Figure 3.28 Narrow Hanover Street was laid out in pre-colonial times and has been attracting people since—its consistently changing and mostly narrow storefronts keep things interesting for pedestrians today (Source: Ann Sussman).

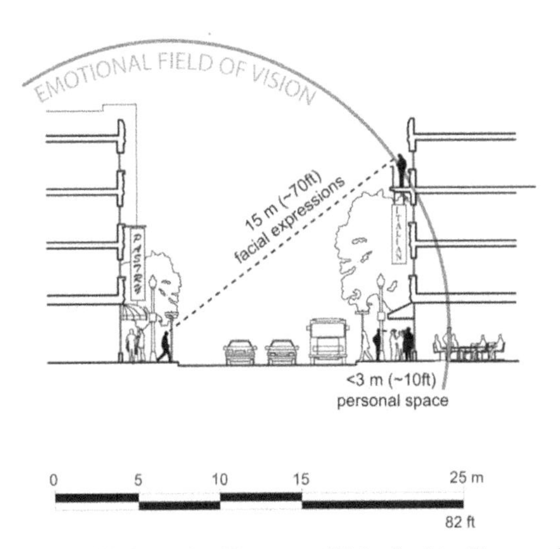

Figure 3.29 At under 20 meters (65 feet) wide, Hanover Street offers us a range of visual and emotional experiences without our having to expend much effort. Its dimensions make it easy to see others, something we innately enjoy (Source: Trey Kirk).

This chapter has suggested how face-processing plays into our sense of aesthetics, via pareidolia or sometimes outright face copying, and helps create buildings and places where our brain subconsciously engages. It does not mean to imply that designers need to draw faces in every building elevation; on the other hand, understanding our face-processing predisposition gives them far greater control over their creative palette and a chance to predict how their work will be perceived by others in a face-obsessed species. Faces, though key, clearly are not the only pattern we have a special affinity toward. Next, we turn to the 'golden rectangle' (see Figure 3.30), which comes up frequently in the history of architecture. In this case, however, we explore it as a biological phenomenon that happens to fit parameters that are set by our face. Chapter 4, Shapes Carry Weight, examines the face's principal attribute, its symmetric shape. Bilateral symmetry predominates in life, and it turns out has significant biological significance that influences architecture, planning, and us.

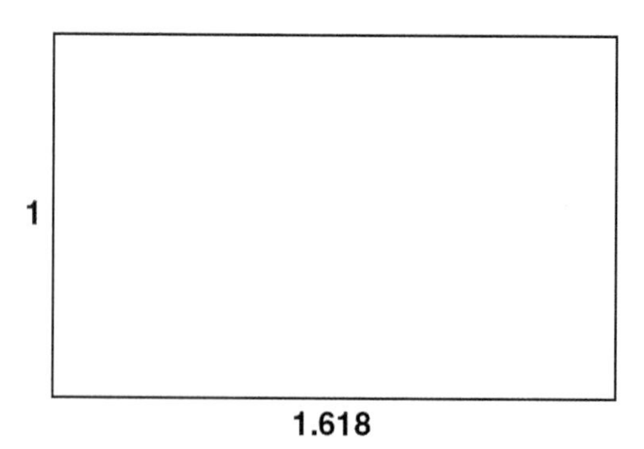

Figure 3.30 The golden rectangle (Source: Trey Kirk).

The 'golden rectangle' is in our face

The golden rectangle, referred to in architectural history since the ancient Greeks, is a rectangle where the length to the height of the shape is roughly in a 3/2 proportion. (The precise relationship is Length/Height [Vertical] equals 1.618, also known as the golden ratio, diagrammed in Figure 3.30) Sometimes revered as having near mystical powers, the shape's defining feature is that when one side is split off into a square, another golden rectangle and square appears *ad infinitum*, as shown in the drawing above (and, drawing an arc within each successive square, creates a golden spiral): (see Figure 3.31).

In architecture, the rectangle can be overlaid on the elevations and/or plans of many of the world's iconic monuments, from the Parthenon in Athens, *c.* fifth century BCE, to Notre

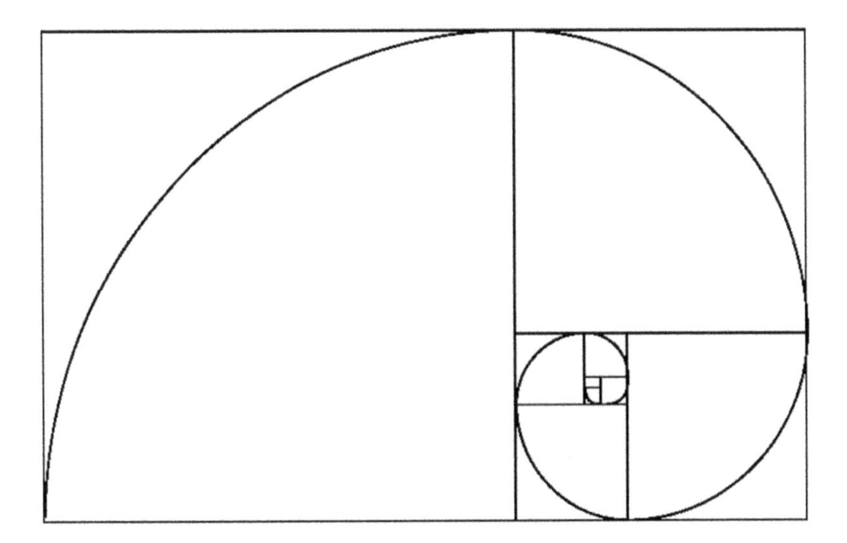

Figure 3.31 The 'golden spiral' forms within the golden rectangle; in a golden rectangle a square cut from the rectangle produces another golden rectangle… *ad infinitum* (Source: Trey Kirk).

Dame in Paris, built from the twelfth through fourteenth century, to Le Corbusier's early twentieth-century villas, although whether it was used in the generation of all buildings where the proportion fits after the fact, is uncertain (see Figure 3.32 and 3.33). The historical record is not always clear. On graphic design websites today the golden rectangle is recommended as a shape to use for the best results in image and photo cropping.

Why is the golden rectangle so common and consistently called out as aesthetically pleasing across so many centuries? It starts with the layout of our eyes and entails the physics of the flow of information from the image to them, according to Adrian Bejan, Duke University professor of mechanical engineering.

Humans have binocular vision. In most cases we see with two eyes and our fields of vision overlaps. In Figure 3.34, each circle represents the approximate area each eye sees, Bejan explains in a 2009 paper, 'The Golden Ratio Predicted: Vision Cognition and Locomotion as a Single Design in Nature.'

Figure 3.34 also illustrates how our vision is predominantly horizontal, which from an evolutionary standpoint again makes sense: danger in man's past generally lurked from the sides, not from the top or bottom. What is quickly apparent, is the

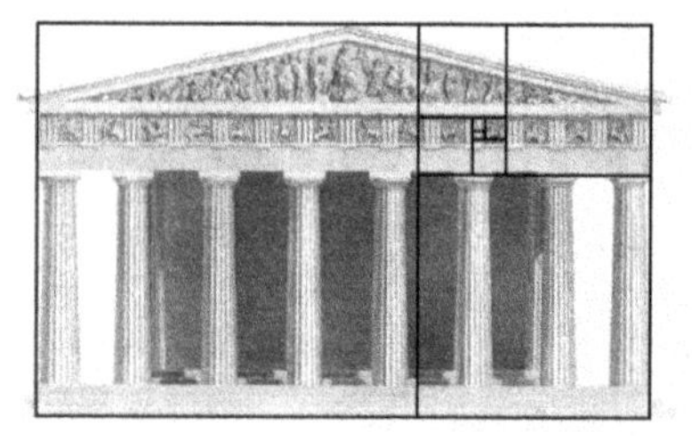

Figure 3.32 and Figure 3.33 Elevations of Le Corbusier's Villa Stein in Garches, France and the Parthenon in Athens fit within the golden rectangle (Source: Trey Kirk).

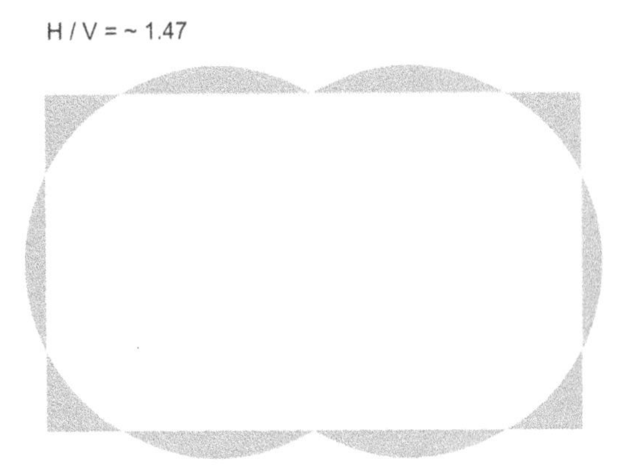

Figure 3.34 The human viewport, evolving for fast horizontal scanning, approximates a golden rectangle (Source: Trey Kirk).

rectangle made by our field of vision approximates the golden rectangle closely at almost the same 3/2 proportion, (more precisely, 1.47 vs. 1.61).

Figure 3.35 illustrates the limits of our visual field. We cannot see very far up without tilting our head, which requires exerting more effort on our part, and our binocular vision does not work at the periphery.

What the above clarifies is how the golden rectangle easily meshes with us, and specifically our viewport and energy-conserving habits. As mentioned earlier, we tend not to do extra work if we do not have to. If you want people to read something quickly—without requiring they exert extra effort—make it that shape. The golden rectangle is like a key that fits the specific lock of our visual field and energy-conserving habit. This can help explain why famous building elevations and familiar objects, from paragraphs in textbooks to standard paper sizes to TV sets to digital displays are close to the 3/2 proportion, as illustrated in Figure 3.36.

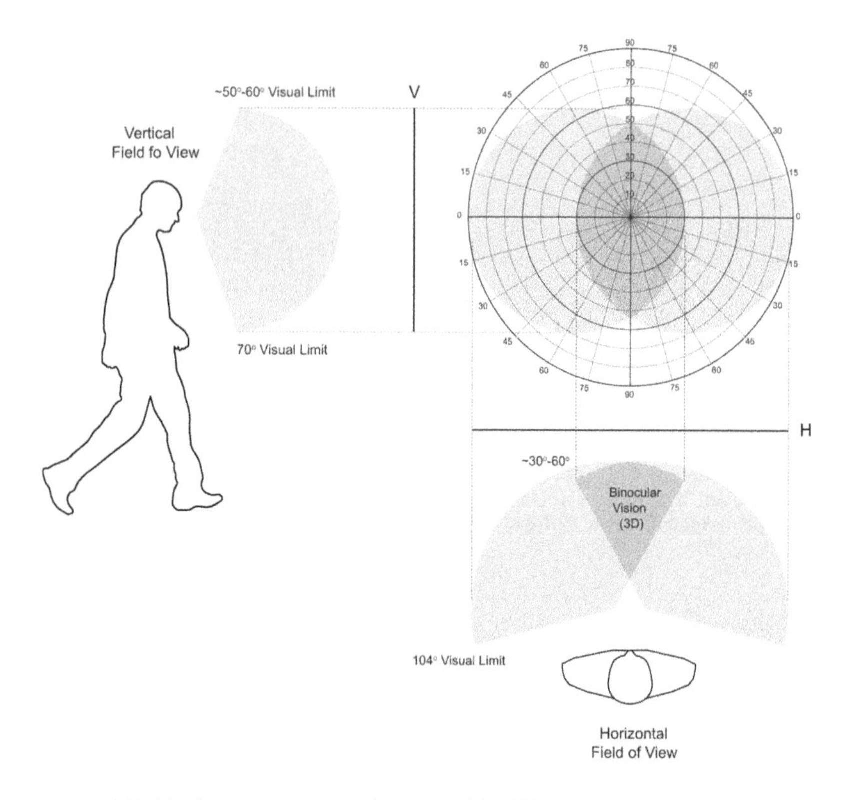

Figure 3.35 Each eye can sweep between 100–120 degrees in the vertical (see diagram above) and horizontal direction. Humans are bifocal (which enables our depth perception), and the horizontal range of both eyes overlap, creating a rectangle where the Horizontal/Vertical (H/V) approximates 1.5. (Source: Trey Kirk).

Bejan, however, takes this golden rectangle analysis a step further, arguing why the shape is efficient for humans to see. When we scan our environment, he explains, different neurons activate depending on the orientation of what is in front of us. Some brain cells specialize in reading lines that are vertical, others fire for horizontal, others when the line is at a slant. Or, as neuroscientist Kandel writes, cells respond 'selectively' depending on linear orientation, (presenting more evidence for

Figure 3.36 The human field of vision superimposed over the golden rectangle shows its relationship to common media dimensions. Things sized this way require less effort to see (Source: Trey Kirk).

how the mammalian eye is not like a camera (Kandel 2012: 26). Bejan points out the horizontal scans we make are faster than the vertical; they must cover the longer horizontal distance in the same time-frame to present us a coherent image of the world. Shape determines how fast an image is 'perceived, understood, and recorded.' What is magical about the golden rectangle? When something has its shape, "the horizontal sweep takes just as long as the vertical sweep." (Bejan 2009: 99) Due to the way we evolved, it happens that the golden rectangle shape requires the minimal scanning time and is maximally efficient.

Moreover, according to Bejan, the proliferation of the golden rectangle follows his 'constructal law,' a phenomenon of physics he defined in 1996. Nature, whether creating a tree with many branches, vascular patterns of a lung, or channels in a river bed, relies on similar patterns to increase the flow of nutrients, air and/or water. Frequent branch-like patterns occur with larger and larger channels leading to major arteries, like in a highway or a biological organ, such as the lung. Rather than viewing this pattern as arbitrary or simply a curiosity, Bejan explains it as a consequence of physics, specifically the 'constructal law,' which he defined as follows:

> For a finite-size system to persist in time (to live), it must evolve in such a way that it provides easier access to the imposed currents that flow through it.
>
> (Bejan 1996: 99)

The 'constructal law' is at work everywhere, he argues. It shows up in the way the brain's neurons, the cells with long fiber-like branches (axons and dendrites) responsible for brain communication, create networks:

> The architecture of the brain consists of bundles and bundles of constantly forming and adjusting tree-shaped flows. The reason is that tree-shaped architectures provide easier and easier point-volume and point-access.
>
> (Bejan 1996: 99)

And it shows up in the fast way our visual system processes the Golden Rectangle:

> This is the best flowing configuration for images from plane to brain, and it manifests itself frequently in human-made shapes that give the impression that they were 'designed' according to the golden ratio.
>
> (Bejan 1996)

Finally, Bejan believes the 'constructal law,' can help us better understand our evolutionary history and in particular the significant interrelationship between the way we see, think, and walk. In an evolutionary time-frame, the eye emerged after animal locomotion, not before, Bejan notes. Better vision in humans permitted our faster, more efficient, and more directed movement. "With vision and cognition the flow of animal mass designs for itself ceaselessly better channels to flow: straighter, safer, with fewer obstacles and predators" (Bejan 1996: 101).

He concludes:

> The punch line of the golden ratio story is unexpectedly much bigger than the golden-ratio prediction itself. It is the oneness of vision, cognition and locomotion as the design of animal mass movement on earth.
>
> (Bejan 1996: 103)

Why is this significant for architecture or planning? Bejan's work suggests the importance of making things in cities consistently visually compelling on the ground floor wherever the plan calls for creating places that appeal to people and encourage their movement forward (see Figure 3.37). Visual stimulation engages our brain and makes us move. Viewing and ambulating evolved interdependently: they are of apiece. Innately we may already know this at some level; Bejan's constructal law grounds the supposition in physics.

Figure 3.37 The 'constructal' law at work: visually compelling windows in Lower Manhattan propel people down the street; the low window frames accommodate the natural tilt of the human head when the body is walking. Our vision and locomotion are interdependent, a consequence of our evolution and a function of biology (Source: Ann Sussman).

Case Study: Society Hill, Philadelphia, Pennsylvania

Philadelphia, the second largest city on the east coast of the U.S. after New York, has the largest collection of eighteenth- and early-nineteenth-century architecture in the country. It is in Society Hill, its central neighborhood, named for the Free Society of Traders, a company that was chartered by the state's founder, William Penn (1644–1718). The neighborhood, part of Penn's original plan for the city, features blocks of brick row houses in the Georgian and Federal style, some still with old cobblestone streets. Adjacent to the docks on the Delaware River and close to historic Independence Hall, where the Declaration of Independence and U.S. Constitution both were adopted, Society Hill was home to Philadelphia's merchant and upper classes for generations. Like many American urban centers, however, the neighborhood fell into decline in the mid-twentieth century with the city's expansion and middle-class flight to the suburbs. It was labeled a slum, considered a shell of its former self, riddled with empty buildings and vacant lots. Targeted for federal urban renewal funding in the late 1950s, Philadelphia's City Planning Commission, under the authoritative leadership of architect Edmund Bacon, selected a development team for the retrofit, including rising architectural star, I. M. Pei (b. 1917). Rather than completely raze the neighborhood as happened in Boston's Scollay Square, and had been a distinct possibility here, the group evolved a distinctively unorthodox hybrid approach. They saved salvageable row houses; the city sold some 600 to buyers inexpensively with strict guidelines on how to restore them; they created dozens of new three-story townhouses, many designed by Pei, to fill-in the vacant parcels in the neighborhood; and lastly, to increase density and broadcast the city center's resurgence from afar, they built three thirty-one-story residential towers, also designed by Pei, in a neighboring park (see Figure 3.38) In the figure-ground drawing (see Figure 3.39) the towers are in the center right of the diagram, and a group of

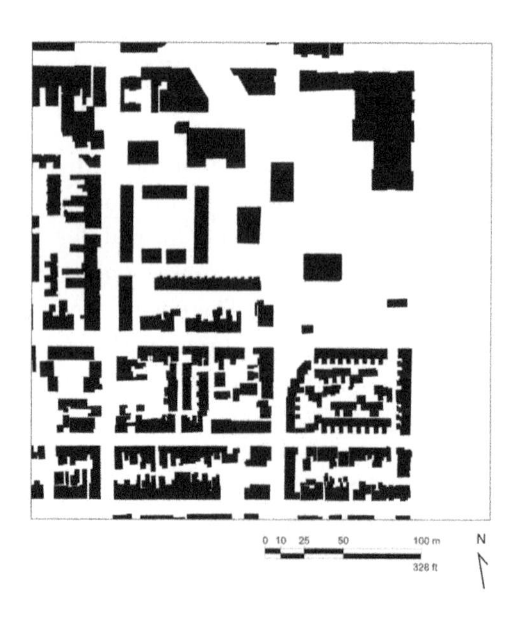

Figure 3.38 Society Hill, Philadelphia. The central city neighborhood has the highest concentration of eighteenth- and early-nineteenth-century architecture in the U.S., and includes three 31-story residential towers, at center right in the figure-ground drawing, built as part of a federally sponsored urban renewal effort completed in 1964 (Source: Nora Shull).

Pei's low-rise townhouses arranged in a large square footprint, are directly to the west (or left).

In the fifty years since construction, history has favored Society Hill's redevelopment approach, but in different ways, revealing urban planning's own transformation. The project, which won a Progressive Architecture (P/A) award in 1961, was called out as exemplary again in 2014, by *Architect*. In an article titled 'Philadelphia Resurgent' the urban renewal team is lauded for cleverly marrying old and new. While the 1960 critics seemed most impressed by the idea of building thirty-one-story apartment buildings in a historic city center, today's critics marvel at something else entirely, the plan's respect of old street patterns and scales. Society Hill's redevelopment:

Figure 3.39 Redeveloped low-rise town houses mitigate the transition to I. M. Pei Towers in Society Hill, Philadelphia, and were constructed as part of the urban renewal project (Source: George Cserna/Avery Architectural and Fine Arts Library Columbia University).

> ...blends sensitively into the pre-existing urban pattern. Dozens of new townhouses designed by Pei echo in both scale and materials the many historic houses that were also restored under the plan. The gaps between these preserved structures became infill sites for sympathetic new construction.
>
> (Dixon 2014)

The new Society Hill kept a lot of the factors, the multiple blocks and intersections that made the neighborhood so successful for walking over generations, it fit the needs of a bipedal forward-looking mammal. Interestingly, the towers, while lauded originally, have come in for criticism in the decades since opening, particularly for how they meet—or fail to meet—the street. Even Pei conceded in later years, in a book reconsidering his work, that the towers had disappointing elements:

> The towers' weakest points were where they met the ground: the arcades formed by the slender columns at the bases were unrelieved by any windows or other inviting detail and as a result were rather chilly spaces.
>
> (Wiseman 1990: 65)

The strength of the recreated town houses, we would argue, is not only that they encourage ambulation because they make for street alignments, or are made of brick, (a factor frequently cited in their popularity), but because with their big windows and doors on and/or at street level, they recreate faces (see Figure 3.40 and 3.41). The faces in the mid-twentieth century row houses are more abstract than in their eighteenth and nineteenth-century iterations, but they are nevertheless there and that impact, as this chapter has tried to convey, can not be understated. (Whether these mid-century elevations contain embedded golden rectangles might also be an interesting avenue of study.)

Figure 3.40 The windows and doors of I. M. Pei Town Houses in Society Hill, designed to repair the old neighborhood fabric, can easily be assembled to make abstract faces (Source: Wikimedia Commons).

Figure 3.41 Eighteenth- and nineteenth-century row houses in Society Hill give the neighborhood its historic charm and can seem face-like (Source: Wikimedia Commons).

Exercise for Chapter 3: Faces and Spaces

- Go to a popular town or city square; get a map or create a figure-ground drawing of the area; is the square within the 'social field of vision' or outside it? Visit a shopping center or mall and explore how the more intimate visual thresholds mentioned above are at work in the store or restaurant plans and their interior layouts.
- Find an iconic house, building, or tower in your area. Can you find visual primitives of the face in its elevations? Do the same thing for continuous building elevations along popular shopping or walking streets. Also note where there is statuary portraying people as part of the building elevation or near the buildings.
- Bejan's theory of 'golden' rectangles explains the prevalence of this shape in human design and links it to the horizontal orientation of our landscape and viewport. How can 'golden rectangles' be used to generate skyscraper design? (Hint: the United Nations Headquarters in New York City provides one example.)

Notes

1 Acclaimed urbanist Holly Whyte (1983) strongly made this point in his popular book and accompanying film, The Social Life of Small Urban Spaces.
2 The extrastriate body area, or EBA, is specialized for perception of human body and its parts.
3 Parahippocampal place area, or PPA, is believed to be specialized for recognizing landscapes or places.

4

Shapes Carry Weight: Bilateral Symmetry, (Hierarchy), Curves, and Complexity

> Eurythmy is beauty and fitness... found when the members of a work are of a height suited to their breadth, of a breadth suited to their length, and, in a word, when they all correspond symmetrically.
>
> Vitruvius (Chapter II, Sec. 3) (*c.* 15–20 BCE)

Shapes carry weight. We do not look out at the world around us as though all things are equal. We have evolved to register and investigate and prefer certain forms over others in fractions of a second. Neuroscientists studying our responses to specific objects and shapes design experiments for subjects to view things in milliseconds. In those brief moments our brain can subconsciously determine whether or not to flee or step forward well before our conscious mind gets into the act. If this were not the case, the theory goes, we could not have survived until now.

In the previous chapter, we looked at faces, one pattern we evolved to process quickly. A key design feature of faces is their bilateral symmetry. In nature, and the world of architecture and design, bilateral symmetry occurs with regularity. Appreciating why this arrangement has such staying power and importance is explored here. To do so, we briefly take a side trip to Anatolia, the central region of Turkey, and, specifically, to the high plains of Cappadocia.

There are fantastic medieval churches hidden in plain sight in a lunar-like landscape made up of unusual limestone formations in this sparsely populated region of Turkey. The stone has eroded over time creating cone-like shapes sometimes called 'fairy castles' (see Figure 4.1). Meticulously carved inside a few of these cones are Byzantine churches that date from the tenth and eleventh century (see Figure 4.2). Those in the small town of Goreme are famed for their bright-colored frescoes, intricate architectural detail, and traditional church plans.

Structurally, these early churches could have taken any shape, yet the religious imperative of creating churches with familiar interiors appears to have been strong. These inside spaces try to mimic those of their conventional masonry cousins

Figure 4.1 The 'fairy castles' of Goreme, a village in Cappadocia, Asia Minor, Turkey (Source: Deniz Gecim).

Figure 4.2 The Interior of Karanlik (Dark) Church, Göreme, Turkey, dates from the end of the twelfth and early thirteenth century. It has a cruciform plan much like a traditional freestanding medieval church and multiple bilaterally symmetric elements. (Author: Karsten Dörre, Source: Wikimedia commons.)

even when this meant carving out free-standing columns that are not actually needed to hold anything up. The cave churches have cruciform layouts and bilaterally symmetric naves leading to a traditional half-rounded apse. Tourists gaze in wonder at the ancient religious architecture so carefully executed inside such remote and unexpected places.

Cappadocia's medieval buildings illustrate how creativity can flourish in adversity, how people can respond creatively when facing challenging topography or limited materials, intent on promoting a cultural and religious vision. Fundamentally, the churches also offer a study in the power of a shape.

The bilateral symmetric form has a history like none other in our built environment and in our own human history. Significantly, the two domains are linked. Looking at bilateral symmetry provides a telling example of how our sense of aesthetics is, at root, biological. Beyond the churches carved into stone in Goreme, bilateral symmetry prevails in traditional architecture elevations and plans around the world: it is in seventeenth-century Chinese temples, the fifth-century BCE Parthenon in Athens, the fifteenth-century Aztec temples in Mexico, and the nineteenth-century Trinity Church in Boston's Back Bay (Figure 4.3).

Across civilizations, the bilaterally symmetric plan and facade is often used to evoke power and convey worldly as well as spiritual might. Sometimes it combines both. For example, the Martha-Mary Chapel, in Sudbury, Massachusetts (see Figure 4.4) was built by the American industrialist Henry Ford in the early 1940s in memory of his mother and mother-in-law. The fact he could afford to build such a shrine obviously reflected his considerable wealth.

Figure 4.3 Trinity Church in Boston's historic Back Bay neighborhood, designed by architect Henry Hobson Richardson, and is a study in the power of bilaterally symmetric shape (Source: Ann Sussman).

Figure 4.4 Martha-Mary Chapel, Sudbury, Massachusetts (Source: Garry D. Harley) built by industrialist Henry Ford to honor his mother and mother-in-law, *c.* 1941.

The form is often connected to showing off wealth. Below is the largest chateau in the Loire Valley, Chateau de Chambord, built for the sixteenth-century king of France, Francois I, as a hunting lodge (see Figure 4.5). Today, it is the most visited tourist attraction in the Valley. (It was never completed, and on careful inspection is not perfectly symmetrical but close.)

Bilateral symmetry is found in much the same way in interiors, such as in the living room of a Dutch Colonial house from the early 1920s, which a wealthy American businessman had built for himself at the age of 27 (see Figure 4.6, the Webster S, Blanchard house, Acton, Massachusetts). What greater way could a young man display his power? The room has a far simpler style, but similar approach, a bilaterally symmetric plan adorned with multiple, repeating bilaterally symmetric shapes.

Figure 4.5 Chambord Castle in the Loire Valley, France, designed as a hunting lodge for the French King Francois 1st (1494–1547) and never completed (Source: Garry Harley). It is the most visited estate in the Loire Valley today.

Figure 4.6 Symmetry conveys power in interior architecture: this craftsman-styled living room, designed as public entertainment space for the Webster S. Blanchard house, was built for a young businessman in 1922 in the outskirts of Boston (Source: Ann Sussman).

Humans are bilaterally symmetrical more or less, as is much of life around us. The link between our form and Classical building tradition is well known and is often examined in architectural texts. Ancient Greeks modeled the columns in their temples directly after the human body, for instance, the capitals representing the head, the shaft the body and the base, the feet. The Roman architect Vitruvius wrote that the architect's work should reflect the body's proportions and its symmetry. His treatise, *De Architectura*, (*c.* 15 BCE) is one of the earliest known texts to link the design of buildings with the architecture of the human body. Leonardo de Vinci famously celebrated the connection in his drawing 'Vitruvian Man,' *c.* 1490, depicting ideal proportions and their basis in the human form and crucial connection to perfect geometries, the circle, and square (see Figure 4.7).

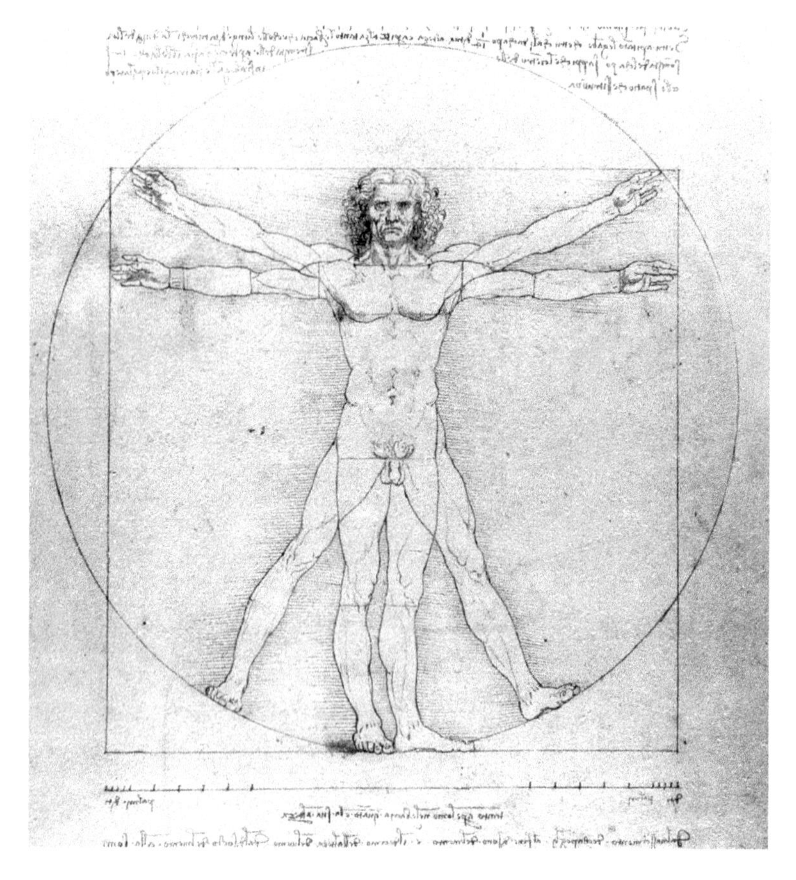

Figure 4.7 Vitruvian Man, by Leonard da Vinci, *c.* 1490, with his text surrounding it, illustrates a Renaissance ideal: man's perfection within nature and embraces the classical notion of nature's perfect geometries. To get this to work, da Vinci ingeniously places the center of the circle in the man's navel and the center point of the square in his genitals (Source: Wikimedia Commons).

"Both of these shapes—the circle and the square—were symbolically important in the design of temples because of their geometric purity," explains the writer Hugh Aldersey-Williams in *Anatomies: A Cultural History of the Human Body* (2013: 26). "It was important to connect them with the human figure in order to demonstrate its divine proportions."

In the Renaissance, sixteenth-century architect Andrea Palladio, influenced by Vitruvius, took this appreciation of bilateral symmetry and divinity literally to new heights and lengths, mandating it in the construction of villas as he explains in his famous architectural treatise, *The Four Books of Architecture* (1570):

> The rooms ought to be distributed on each side of the entry and hall: and it is to be observed that those on the right correspond to those on the left, that so the fabrick may be the same in one place as in the other...
>
> (Palladio [reprint 1965]: 27)

Palladio practiced what he preached, as can be seen in his plans for his most famous architectural legacy, the Villa Almerico Capra, also known as 'La Rotonda,' begun in 1566 and completed in 1585, five years after his death (Figure 4.8). In 1994, it became part of a UNESCO-designated World Heritage Site.

Bilateral Symmetry and Biology

Bilateral symmetry in plans and architecture today has become common to the point that it may seem predictable, tedious, or something to avoid. From the biological standpoint, however, the shape itself is anything but. Without bilateral symmetry, biologists note, the possibility of our existence as human beings is moot. Moreover, the science suggests bilateral symmetry has within it key efficiencies that help us navigate our world, both animate and inanimate. It turns out that it is not at all coincidental that much of the architecture depicted above uses bilateral symmetry to evoke power, prestige, and might.

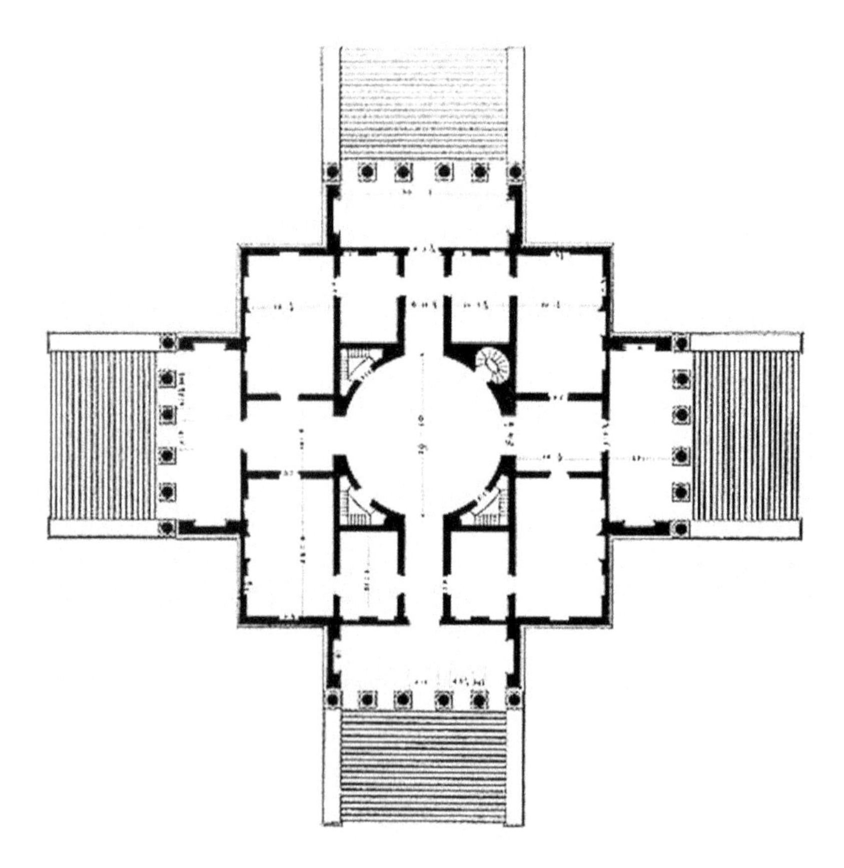

Figure 4.8 Villa Capra, 'La Rotonda', Vicenza, 1566, by Andrea Palladio (1508–1580) from Planta de "i quattri libri," (1570). In the hands of the high Renaissance master Andrea Palladio, symmetric plans and elevations reach an apotheosis. Publicacion de Ottavio Bertotti Scamozzi, 1778 (Source: Wikimedia Commons).

In the Beginning...

Early life forms were not necessarily bilaterally symmetrical. Sea sponges, some 650 million years old as a species, root to the ocean floor, do not have a brain or circulatory system, and are asymmetrical (Figure 4.9). But following the Cambrian Explosion, 540 million years ago, where there is rapid appearance of multiple

Figure 4.9 Asymmetrical pink lumpy sponge (Author: Nick Hobgood, Source: Wikimedia Commons,). Sponges are considered the foundational species for all life that followed including our own. Recently pushed back in age, they are now thought to have emerged 550 to 750 million years ago.

species in the fossil record, bilateral life dominates. Today, "99% of modern animals are members of the evolutionary group Bilateria," (Finnerty *et al.* 2004: 1335) biologists note.

Bilateral symmetry conveys significant advantages for its species, including us. The form is ranked as "an important advance" because "it opened the way for the development of directed motion, improved organs of sense and, eventually, the enlarged and highly complex mammalian brain" (Prosser 2012).

In bilaterally symmetric organisms, an axis, in humans, the vertical, has halves that are approximate mirror images. Things naturally come to a head in this geometry, the way they do not in an asymmetric or other arrangement. This permits movement in one direction and the centralization of the nervous system in one place, a brain. Promoting directional locomotion, this form of symmetry also encourages the development of vision and cognition interdependently. It is intriguing to hold on to the idea that all of our three key abilities (vision, cognition, locomotion), are interconnected (also mentioned in Chapter 3) and a consequence of a bilateral symmetric layout.

Bilateral symmetry not only influences the way we walk or how we see, but also appears to deeply connect to our emotions, and inform what we like and find attractive in people and other animate and inanimate things in the world. Biologists refer to this tendency as "an evolved preference" (Cardenas and Harris 2006: 11). Here is how Charles Darwin relates our bias toward symmetry in humans and other species in *The Descent of Man and Selection in Relation to Sex*, his second book on evolutionary theory, (published in 1882).

> . . . the eye prefers symmetry or figures with some regular recurrence. Patterns of this kind are employed by even the lowest savages as ornaments; and they have been developed through sexual selection for the adornment of some male animals.
>
> (Darwin 1882: 93)

More recently, psychologists have tried to tease apart the extent preferences for symmetry appear to be hardwired in us, independent of our cultural or 'savage' heritage. Recent psychology studies, for instance, have explored whether adding symmetrical patterns to faces and craft objects enhances their appeal. They do, the studies reported: people consistently prefer symmetrical additions over the asymmetric (Cardenas and Harris, 2006). Researchers also note symmetrical patterns prevail in arts and crafts cross-culturally, whether it is in pottery, fabric design, tile ornamentation, or body decoration. Significantly, the tendency in crafts also seems to have arisen independently throughout the world, suggesting its primal place for our species. Bilateral symmetric objects are found in diverse, far-flung regions, ranging from the Navajo in the American West, to the Aonikenk, tribes of Patagonia, South America, to the Yoruba tribe of Nigeria (Cardenas and Harris 2006) (see Figure 4.10).

There seems to be a persistent tendency in this research to link how we see faces to how we see other things (an instance, again, of the significance of faces noted in Chapter 3). Men and women find symmetrical faces more attractive than non-symmetrical ones, research has confirmed. "Preference for

Figure 4.10 Bilaterally symmetric rams appear in Antioch Culture, House of Ram's Heads Floor Mosaic (detail), late fifth or early sixth century CE, marble and limestone tesserae, 76.2 x 208.3 cm, Worcester Art Museum, Worcester, Massachusetts, Excavation at Antioch, 1936.33 (photo credit: Ann Sussman).

symmetry, while perhaps acquirable through cultural processes, is rooted in our evolutionary history," a recent paper in the field noted (Cardenas and Harris 2006: 3).

One frequent explanation for the symmetry preference is the "good genes" hypothesis, which holds that more symmetrical faces were and still are seen as healthier and fitter, at a glance signifying resiliency in a mate. A symmetrical face and body advertises its ability to withstand and overcome the vicissitudes of life, and hence, is more likely to reproduce. In a study reported in the journal *Evolution and Human Behavior* in 2006, "Symmetrical decorations enhance the attractiveness of faces and abstract designs" psychologists Rodrigo Andres Cardenas and Lauren Julius Harris at Michigan State University gave forty undergraduate students a series of symmetrical and asymmetrical patterns selected from different indigenous cultures (Cardenas and Harris 2006). They told them to pick the ones they preferred, the students consistently chose the symmetrical pattern (see Figure 4.11). Given designs that were symmetric around the Y-axis versus those that were asymmetric, they again tended to pick symmetry around the vertical axis as preferable.

The authors concluded, "the preference for symmetry extends to the cultural products of facial paint and the decorative arts." But, they caution, "although symmetrical art is very common, the preference for symmetrical facial features is more likely to be constant than the preference for symmetrical art" (see Figure 4.12) (Cardenas and Harris 2006: 16). They also hypothesize why the preference prevails: "One possibility is that the adaptive value of detecting symmetry in potential mates generalizes to other objects" (ibid: 16).

Perhaps because we like looking at faces and have evolved to take in and emotionally read them quickly, we also favor the main facial attribute, bilateral symmetry, in things we make and place around us. Faces ground and orient us in a random world from infancy onward. One might hypothesize there is a certain efficiency and predictability to designing buildings that reflect

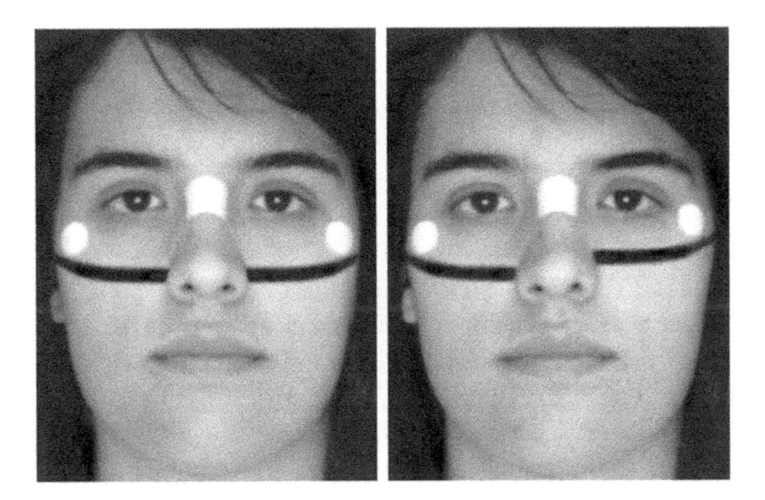

Figure 4.11 Photographs like the ones above were used in psychological research to show the innate symmetric preferences in humans. The research found that test subjects preferred a symmetrical face (left) with added symmetric decorations over an asymmetric face (right) applied with asymmetric paint (Source: Rodrigo A. Cardenas).

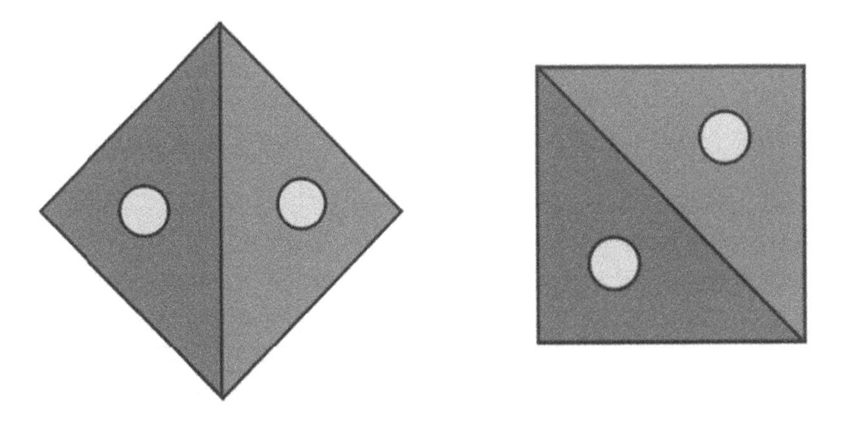

Figure 4.12 In psychological tests, subjects consistently picked a pattern symmetrical around a vertical axis as more attractive than one symmetrical about a non-vertical axis (Source: Rodrigo A. Cardenas).

this arrangement not only because we are predisposed to take the form in, but because such new constructions may more likely reassure us, too (see Chapter 3 for more details).

Psychologists studying symmetry perception have found that people process vertical bilateral symmetry (oriented around the vertical axis), in objects more quickly than other forms of repetition or symmetry (Makin *et al.* 2012: 3250). One theory for the vertical bias is that it is built-in, an artifact of the way our eyes sit in our head, parallel to the horizontal plane. (Perhaps, if our eyes were perpendicular to the horizon our bias would be in that direction.)

Researchers have also learned that looking at symmetrical objects subconsciously activates our smiling muscles more than looking at random patterns (Makin *et al.* 2012: 3255). And when we smile, we are more likely to feel calm or reassured. "We have a built-in aesthetic preference for symmetry," surmises neuroscientist VS Ramachandran, in *The Science of Art, A Neurological Theory of Aesthetic Experience*, his 1999 paper with William Hirstein. The authors label symmetry one of the eight theoretical 'laws' of aesthetic experience (Ramachandran and Hirstein 1999). They suggest the key reason the shape resonates is because of its survival link: "Since most biologically important objects – such as predator, prey or mate are symmetrical, it may serve as an early-warning system to grab our attention to facilitate further processing of the symmetrical entity until it is fully recognised" (Ramachandran and Hirstein 1999: 27).

Studies also show that symmetric objects have redundancy inherent in their design that appears to contribute to faster mental processing of the form. It appears once we've read half of a symmetrical shape, our mind has predicted the other. Finally, no discussion of symmetry is complete without mentioning how historically it is consistently linked to beauty as well as order and organization. "Beauty is bound up with symmetry," the German mathematician and philosopher Herman Weyl (1952) summarized in his book, *Symmetry*.

"Symmetry, as wide or as narrow as you may define its meaning, is one idea by which man through the ages has tried to comprehend and create order, beauty and perfection" (Weyl 1952: 5). From a biological standpoint we can surmise that beauty is connected to symmetry for one good reason mentioned in earlier chapters, it is bound up and cannot be teased apart from survival.

Curves

In terms of innate preference for shape, humans also have a clear bias for curves over straight or sharp lines. Aesthetic judgments are a complex matter engaging many part of the brain. Studies in the field of aesthetics more than a century ago found that when it comes to 2D and 3D objects, curves elicit feelings of happiness and elation, while jagged and sharp forms, tend to connect to feelings of pain and sadness (see Figure 4.14).

"Curves are in general felt to be more beautiful than straight lines. They are more graceful and pliable, and avoid the harshness of some straight lines," psychologist Kate Gordon wrote in her book *Esthetics*, published in 1909 (Gordon: 169). Even "the most simple abstract line... may have an emotional effect and meaning of its own. (Gordon: 160)."

Numerous psychology research papers have documented these findings since. Measuring student responses to angular versus curved typefaces in 1968, psychologist A. J. Kastl found feelings such as "sprightly, sparkling, dreamy and soary," arose viewing curving fonts, while moods of sadness went with "angular, bold and perhaps serif type." Similarly, when psychologist Rudolf Arnheim (1969) asked students to describe a "good and bad marriage" using a simple line drawing alone, he found that a continuous undulating smooth curve was seen as representing the loving union, and an irregular spiky line the bad one (Arnheim: 125).

Hierarchy helps

Many bilaterally symmetric forms also boast clear hierarchy. They have a top, middle, and bottom—frequently tripartite. If symmetry suggests organization and intentionality; hierarchy does too. We see this in the human form, with its head, body, and feet; the human face, with its eyes, nose and mouth. In traditional architecture cross-culturally, the tripartite hierarchy also tends to dominate on building elevations. It suggests an order and intention we can easily and intuitively understand. Perhaps the fact we have such familiarity with the arrangement, and are built to interpret it wordlessly, accounts for its frequency. A building with a clear roofline, middle section, and articulated base looks complete, resolved, familiar, much like an articulated or simply rendered figure or face (see Figure 4.13).

Figure 4.13 The Taj Mahal (*c.* 1653) in Agra, India has a clear, hierarchical shape with a tri-partite arrangement, a top, middle, and bottom not unlike a face; it can also be seen, not incidentally, as a study in curves, another form humans innately prefer (Author: By Yann, Source: Wikimedia Commons).

Figure 4.14 Dancing Maenad, Roman, *c.* 27 BCE–14 CE, Metropolitan Museum of Art, portrays a mythical dancing devotee of the Roman god of wine, Bacchus, and can be viewed as a study in the appeal of curvaceous shapes (Image copyright © The Metropolitan Museum of Art, Image source: Art Resource, NY).

The curve bias carries over to our reaction to art and can even affect patient feelings and recovery rates according to hospital design expert, Robert Ulrich. In one study of 160 intensive-care patients in a Swedish hospital in 1993, he monitored the patient responses to six different views: two of nature, two of abstract art, and two of a blank wall. "... A rather surprising finding was that an abstract picture dominated by rectilinear forms produced higher patient anxiety than control conditions of no picture at all," Ulrich wrote (2002: 7).

The impact of our preference for curves transfers to our feelings about architecture according to a more recent study using fMRI, psychologist Oshin Vartanian and colleagues (2013) determined. In this test, twenty-four subjects were given 200 pictures of architectural spaces to look at. Half of the images were rectilinear; half curvilinear. "As predicted, participants were more likely to judge spaces as beautiful if they were curvilinear than rectilinear," the report said (Vartanian *et al.* 2013: 1). "The results suggest that the well-established effect of contour on aesthetic preference can be extended to architecture. Furthermore, the combination of our behavioral and neural evidence underscores the role of emotion in our preference for curvilinear objects in this domain." (Vartanian *et al.* 2013: 1). We like things plump and round.

Why does the curve-bias exist? Humans evolved to assess their environment quickly. Pointed shapes, such as barbs, thorns, quills, sharp teeth, were ever-present threats in our evolutionary past, so it was advantageous to sense them fast and be able to flee if one had to. Psychologically, part of our brain still feels a lion could be at the gate, even as we sit in the living room of a high-rise or suburban tract house. We evolved for this past environment, and are still designed for it whether or not it exists in our present.

"Humans like sharp angled objects significantly less than they like objects with a curved contour," wrote neuroscientists Moshe Bar and Maital Neta in a 2007 paper summarizing a study

Case Study: The Oval Office, The White House, Washington D.C.

Our bias toward curves suggests why the design of the 'Oval Office,' the traditional seat of power of the American president, may carry a psychological advantage for its occupant; not only is it bilaterally symmetrical with the desk centered on its longer axis, indicative of psychological power, but its rounded walls are of an innately preferred shape (see Figure 4.15).

Figure 4.15 Curves define the character and help magnify the power of the Oval Office, seen here with President Barack Obama in 2013 (Author: Pete Souza, Source: Wikimedia Commons).

that scanned participants as they observed more than 200 different shapes (p. 2300). "This bias can stem from an increased sense of threat and danger conveyed by these sharp visual elements." The researchers noted an area of the brain engaged in the fearful responses, the amygdala (sometimes called our 'lizard brain'), "shows significantly more activation for the sharp-angled objects compared with their curved counterparts." They also proposed "that the danger conveyed by the sharp-angle stimuli was relatively implicit." It appears, then, we do not even have to learn much about some things, part of our brain is set up to have us run from a sharp shape. (For more on the human brain morphology and function, see the Appendix.)

"We are still innately drawn to settings whose characteristics hold some survival advantage," wrote Grant Hildebrand (2008), architect and University of Washington professor emeritus "[E]ven though that survival advantage may no longer have any practical value for us" (p. 263). Our present subconscious responses reflect a past trajectory that we can not rewrite.

Order and Complexity

Favoring curves and symmetry, we also enjoy complexity. "Order alone is monotony," Hildebrand wrote in a chapter in *Biophilic Design*, "complexity alone is chaos." (Hildebrand 2008: 264). He explains how this architectural attribute asserts itself in other arts, including music and dance:

> ... there is substantial empirical evidence that we are genetically programmed to respond positively to complexly ordered sound (music) but not to chaotically complex sound (noise). One might argue, similarly, that consciously or unconsciously, we distinguish architecture from "just building" by the evident order and complexity of its materials and spaces.
>
> (p. 265)

Architecture's connection to music was perhaps most famously penned by German writer Johann Wolfgang von Goethe (1749–1832) who wrote: "Music is liquid architecture; Architecture is frozen music." We by design enjoy processing complex aural and visual stimuli, when presented with an order inherent.

Fractals in landscape, art, and architecture

Evidence of nature's own embrace of order and complexity is evident in fractals, recursive patterns that occur repetitively in smaller and smaller scales and abound in nature. We find them in snowflakes, leaves, coastlines, and vegetable patterns, such as in the beautiful shape of the cauliflower, a romanesco calabrese, shown in the photograph in Figure 4.16.

Fractals appeal to humans innately according to Richard P. Taylor, a University of Oregon physicist who studies them extensively, and has investigated the human response to the patterns for more than a decade. Taylor's research found that humans appear to have innate preferences for fractal patterns that are not too dense, but not too sparse either. Patterns in the ideal range "generated the maximal alpha response in the frontal region, consistent with the hypothesis that they are most relaxing," he explained (Taylor *et al.* 2011: 18). This also happens to be the range of the particular pattern of trees common to the African savanna, suggesting how our biology and aesthetics co-evolved and became intertwined. (The importance of the savanna landscape and specifically acacia trees for aesthetics are outlined further in Chapter 6, Nature is our Context.)

Taylor has also studied the appeal and power of fractals in art and architecture. The fractal patterns embedded in the work of modern abstract painter Jackson Pollock (1912–1956), for example, who is renowned for his drip-paintings on large canvases,

Figure 4.16 Fractal beauty in a romanesco calabrese cauliflower. Self-replicating forms fascinate us: our interest in these patterns has been linked to our evolution and observation of similar arrangements, such as tree patterns, in nature over eons (Author: AVM, Source: Wikimedia Commons).

may be part of the reason the pieces today fetch prices up to $200 million each, Taylor explains. Our response to fractals is physiological and we like it. Fractals have also been found in famous architecture, including in the design and decoration of medieval Gothic cathedrals, the Eiffel Tower (the landmark in Paris, created by French architect and engineer Gustave Eiffel [1832–1923]), and in Frank Lloyd Wright's (1867–1959) later works including the Palmer House, built in Michigan in the 1950s. Since fractal patterns can induce relaxation, Taylor believes their further study can not only help us learn more about human perception but contribute to improving our built environment. He and his co-authors of the paper, "Perceptual and Physiological Responses to Jackson Pollock's Fractals" (2011) conclude:

> Scientific experiments might appear to be a highly unusual tool for judging art. However, our preliminary experiments provide a fascinating insight into the impact that art might have on the perceptual, physiological and neurological condition of the observer. Our future investigations will explore the possibility of incorporating fractal art into the interior and exterior of buildings, in order to adapt the visual characteristics of artificial environments to the positive responses.
>
> (Taylor *et al.* 2011: 19)

Shapes carry weight and the more we appreciate our response to specific ones, including fractals, the better our designs will be.

Exercise for Chapter 4: Shapes Carry Weight

- Look for symmetry, curves, and complexity in a favorite landmark and/or building elevation. Do these traits show up only in plan, elevation, or both?
- Trinity Church, in Figure 4.3, has elaborate elevations that draw visitors' attention and seem to engage viewers more than any of the neighboring structures viewed in the photograph. Analyze why this may be the case using the concepts presented above.
- Designers claim that intentionally asymmetric arrangements, such as of windows on a building facade, can draw increased attention to the structure. Why might this be the case?
- People evolved to see other people. As mentioned in Chapter 3, we perceive significantly more information as we come closer to another person, the approach has increasing psychological impact. Alone or with a partner, find a building designed to do the same thing. Record your responses—visual and emotional. As a contrast, find a building that presents as a blank and portrays less information the closer you get: record its impact on your emotions and state of mind.

5

Storytelling is Key: We're Wired for Narrative

> On Narrative: "We came to see that the search for attachment—to a person, an object, a work of art, an idea—held open the possibility of feeling not alone… of knowing the meaning of expansive connection between self and world."
>
> Kay Young (2010), *Imaging minds. The Neuro-Aesthetics of Austen,* Eliot, and Hardy: ix

Just as the human brain is primed to find faces, favors symmetry, and enjoys ordered complexity, it runs on narrative. How we see our world and how we see ourselves ultimately involves a story. Narrative is the unusual ability of the mind to create stories and, in the process, find multiple ways of linking to the environment and securing a place in it. Biologists consider our adeptness at coming up with stories highly adaptive. While we share many other traits with other animals, whether it is thigmotaxis, as in Chapter 2, or a symmetry-bias discussed in Chapter 4, no other creature has the capacity to create and continually elaborate its own story. Narrative makes us human, wrote Roland Barthes, the French philosopher and literary theorist (as quoted in Young and Saver 1979):

> (N)arrative is present in every age, in every place, in every society; it begins with the very history of mankind and nowhere has there ever been a people without narrative… (it) is

> international, transhistorical, transcultural: it is simply there, like life itself.
>
> (Barthes, as quoted in Young and Saver 2001: 79)

Narrative not only tells our mind stories, it tells us something about the organization of our mind. In *The Neurology of Narrative* (2001), Kay Young, PhD, and Jeffrey Saver, MD, Professors of English and neurology at the University of California, Santa Barbara, respectively, describe narrative as "the inescapable frame of human existence" (Young and Saver 2001: 79). They note how diverse thinkers over two thousand years, from Aristotle to Barthes, have deduced the centrality of narrative to human cognition without actually knowing the biology or neuroscience behind it. Now, that is changing. Researchers recently have identified the distributed neural network that creates narrative in the human central nervous system.[1] The work shows how these pathways are critical for story-making and for something more critical: our sense of self. An old adage goes: you are what you think. Less elegantly, but more accurately, we might now say, you *are* because of the way you are enabled to create, remake, and remember stories.

Studies frequently look at people with brain damage to better determine what specific regions of the brain do. As a result of certain injuries, for example, someone may lose her ability to speak or see. Even if struck blind or dumb, a subject will remain "recognizably the same person," the authors note. On the other hand, when individuals sustain damage to the neural network involved in story-making, they lose "the ability to construct narrative... (and) have lost their selves."

"Narrative is deeply human." It is the mind's organizing force. "To desire narrative reflects a kind of fundamental desire for life and self that finds its source in our neurological make up" (Young and Saver 2001: 80). It turns out Roland Barthes' thinking would prefigure the scientific findings by decades; he was right, narrative is the dynamic, living process that gives us ourselves.

Imagining scenarios or stories and not actually acting on them is a significant attribute of the human narrative capacity. The term for this behavior is "decoupling," or "the separation of mental action from physical action" (Young and Saver 2001: 82). Biologists again consider it highly advantageous. Decoupling allows us to imagine multiple narratives without "engaging the motor apparatus" (p. 82); its existence has a huge role in allowing us to lead rich and diverse lives. Decoupling permits the creation of imaginative work which makes possible the foundation for the arts. Or as Young and Saver write, research suggests "...the evolutionary origin of the human abilities to imagine literature and the arts may be traced to this functionally advantageous capacity" (p. 82).

Why does this matter for architecture or planning? It suggests one more way people consistently look for orientation and connections to their environment. Much as we seek out faces from infancy on, we look for ways to make attachments and derive meaning from our physical surroundings. Every plan and urban design has the potential to acknowledge and respond to this trait in some way or another, or as is frequently the case in built environments today, ignore it. One could make the argument that it is the inherent lack of a narrative quality in many of the post-war American suburbs, (as opposed to the earlier nineteenth-century street-car versions) that gives these areas their feelings of placelessness and anomie. (A case study at the end of this chapter looks at how one group of suburban residents attempted to address the 'placeless' problem at their town's entry point.)

Narrative can be addressed in different ways: people connect to historical events and figures linked with a location. George Washington's home in Mount Vernon, Virginia, for instance, remains "the most popular historic estate," to visit in the United States, according to the non-profit that runs it (Mount Vernon 2014). People travel there to develop a greater understanding of the nation's first president and his time. Narrative can also be embedded wordlessly, expressed in the spatial sequencing of a plan, for example, including its room layout, orientation, and size.

Some architects' plans are famous for this type of narrative quality, and appear to have been intentionally made to increase our ability to connect, both to nature outside and internally, to our narratively inclined mind.

Frank Lloyd Wright's house plans, for example, are known for their clear sequencing: a small entryway with a low ceiling leads to a tight anteroom and then crescendos in the large public living space with high ceiling, a fireplace ('the hearth'), and a broad view of nature outside (see Figure 5.1). Wright could have designed the homes for residents to walk straight into the main living space but never did. The careful ordering of spaces instead gives the house a story-like flow, magnifying a sense of arrival in the largest rooms and celebrating the home as a significant, dignified, place. By acknowledging our needs for both internal and external connections, Wright's plans ennoble beholder and occupant alike (not surprisingly, he is frequently labeled a 'romantic').

Figure 5.1 The drawings for Frank Lloyd Wright's 'A Fireproof House for $5,000', were published in the *Ladies Home Journal* magazine in April 1907, with an accompanying article written by the architect. The building was expressly intended to be affordable for the American middle class. (Author: Frank Lloyd Wright, http://www.stockmanhouse.org/lhj.html, April 1907, Source: Wikimedia Commons).

The Wright plan in Figure 5.2 first appeared in the *Ladies Home Journal* in 1907. In an article entitled 'A Fireproof House for $5,000,' Wright explains the design is for the "average homemaker" looking for an "inexpensive" alternative to the "overtrimmed boxes" of the time. Wright presents this house not as a palace for the rich, though it may seem high-end, but a home for every man or woman. It uses the up-to-date construction techniques of the time (cast-in-place concrete), and does away with superfluous rooms like butler's pantries and attics to minimize cost and maintenance, he wrote. The first floor, with a ceremonial trellis-covered walkway leading to the front door and its tight entry, makes its big statement upon arrival at the hearth centrally located in the building's forty-two square meter (450 square feet) living room. Wright even specifies how the house should orient to the street, with a principal view illustrated in the drawing above. We cannot help but note how face-like the elevations are and how Wright made do with unusual asymmetrical arrangements of windows in the second-floor bedrooms to maintain the outside symmetry. The facades are vertically and bilaterally symmetrical and designed with a distinct top, middle, and bottom. The roof overhang looks a bit like a hat. There is a narrative quality to the entire exterior of the building; it can easily be read as a face, casting a steady gaze to passersby on the street.

Centuries ago, the idea that narrative has a central place in creating memorable landscape was familiar to architects and garden planners. Designed in the Renaissance, the Villa Lante, fifty miles north of Rome in central Italy, is "regarded by most authorities as the finest of all Cinquecento villas," (Newton 1971: 99) and is essentially a garden folly that uses the biblical story of Noah's Ark and the Flood to frame its layout. Conceived originally as a social playground for Italian clerics and their peers, its recreational value has endured for more than four hundred years. In 2011 it was voted the "most beautiful park in Italy" (see Figures 5.3–5.4). If Frank Lloyd Wright's use of

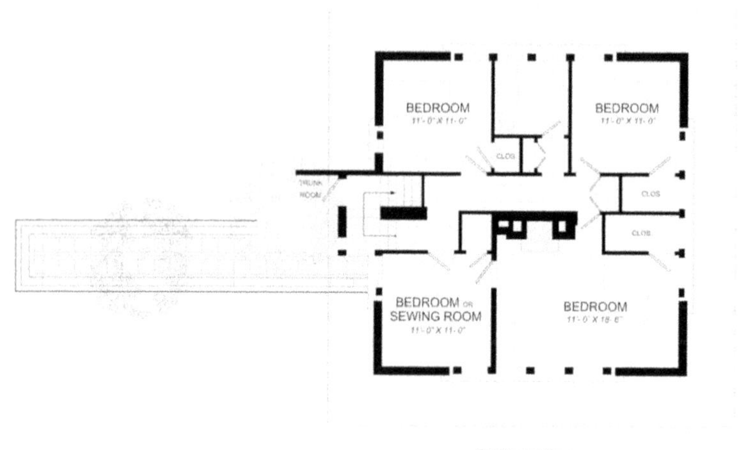

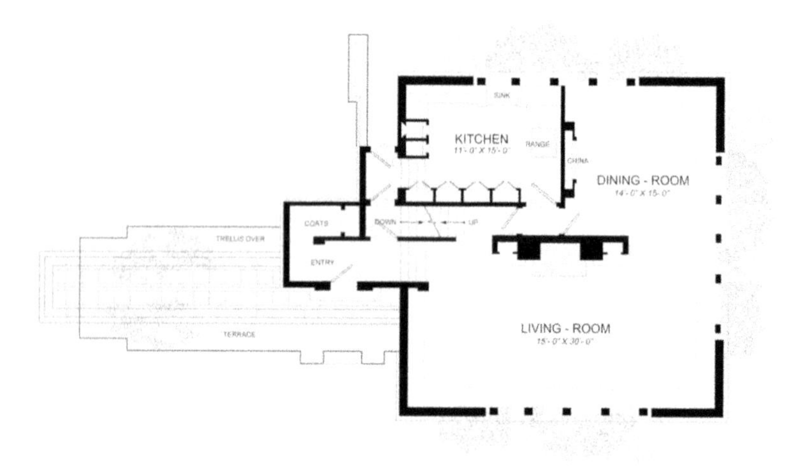

The First-Story Plan

Figure 5.2 Top: Fireproof House, first floor plan; bottom: second floor plan (Author: Frank Lloyd Wright, "A Fireproof House for $5,000", April 2007, Source: Wikimedia Commons).

spatial sequencing is prosaic, Villa Lante's illustrates the dramatic.

Figure 5.3 The Pegasus Fountain at the entry to the sixteenth-century Villa Lante in Bagnaia, central Italy, represents Biblical paradise, a time before man's fall from grace where there was natural abundance on earth (Source: Tom Toft, 'Own Work,' Wikimedia Commons, May 11, 2007).

Figure 5.4 The 'garden finale' at Villa Lante, suggests a new age of hope dawning after 'the fall,' where man's creativity and knowledge can be put to use offering hope and salvation (Source: Robert Ferrari, 'Giardano #2,' Wikimedia Commons, April 26, 2007).

Figure 5.5 The plan of Villa Lante; the principal entry to the garden starts at the top (or south) in the diagram and follows a linear progression down a grade to the exit at the north. The garden sequencing tells the story of man's biblical fall from grace and re-emergence into a world of rationality and hope, represented by a grand fountain centered in a large square of symmetrically arranged plantings at its base (Source: Trey Kirk with permission).

Case Study: Villa Lante, Bagnaia, Italy

Villa Lante was begun in 1511 and completed around 1566. It is famous for its formal plantings on a hillside and an overall plan that steps down four levels, each with a distinct personality. Bilaterally symmetric with a watercourse running down its center and embellished with dramatic fountains, the villa is in the Mannerist tradition, a Renaissance style known for a certain playfulness, and in this instance, particularly in its final sequence exploring the power of Palladian squares and circles laid out on a grid (see Figure 5.5).

"For most observers, what makes the Villa Lante such a compelling experience is probably the handling of spaces in a wonderfully comfortable rhythmic sequence from level to level," wrote landscape architect Norman T Newton in *Design on the Land: The Development of Landscape Architecture* (Newton 1971: 106).

Here, visitors seem to agree: the linear layout of surprising spaces makes the place—as does its underlying biblical story. The experience of walking through the garden was significant enough to be documented in Pope Gregory XIII's 1578 visit. (Lazzaro-Bruno 1977: 555) Today visitors walk along essentially the same path. First, "(t)he Fountain of Parnassus at the entrance to the park identifies the whole villa as a place of contemplation under the inspiration of nature, and also as the ideal realm, the earthly paradise which Parnassus is as well," explains Cornell art historian, Claudia Lazzaro-Bruno. Then, as the visitor passes through the gardens, he or she comes to the Classical myths of the Golden Age, including the Fountain of Acorns, representing the Garden of Eden, "the time when men ate only acorns and honey" (p. 555). But things sour quickly during this Golden Age "during which man became increasingly evil until finally Jupiter decided to destroy him completely by means of a flood (Lazarro-Bruno 1977: 556)." The flood itself happens as the visitor proceeds to the formal garden, at the

'Grotto of the Flood." With the end of the flood, a new Age of Jupiter dawns and it is expressed through the symmetrical square garden replete with squares and circles (see Figure 5.4).

"The serial idea builds upon the sense of progression that often characterizes the experience of the linear space," Alexander Purves, Yale architecture professor emeritus, concludes in *The Persistence of Formal Patterns*.

> The essence of the idea is incremental change. The watercourse at the Villa Lante in Bagnaia bursts from a spring in the side of a hill and drops from terrace to terrace through a series of delightful hurdles until it finally comes to rest in a formal pool. The journey of the water from high to low, from natural spring to artificial basin, parallels a transformation of the landscape from artful nature to rational geometry.
>
> (Purves 1982: 154)

The study of Villa Lante suggests how narrative, in this case orchestrated in carefully rendered steps, can increase people's ability to enjoy place. On the other hand, randomness and a lack of serial progression can lead to discomfort and feelings of placelessness, particularly from the pedestrian perspective. An entry intersection to a post-war suburb of Boston, Massachusetts, offers an example of the diametrically opposite experience to Villa Lante and is explored next.

Case Study: Kelley's Corner, Acton, Massachusetts

Acton, Massachusetts is a mid-sized town, 25 miles northwest of Boston, which tripled in size from 1960 to 2000, a relative late-bloomer in American post-war suburban development (see Figures 5.6 and 5.7). Until the twentieth century predominantly agricultural, it is a bedroom community of mostly commuters now, known for its public schools and for wrestling with how to best manage new growth (von Hoffman 2010). Not unlike many metropolitan towns outside of Boston's inner ring-road (Route 128) the town has more registered vehicles (cars, trucks, SUVs, etc.) than actual residents (Sussman 2011). Another telltale sign of its recent rapid development is its lack of a town center with stores and cafes. The entry intersection off the highway greets visitors instead with acres of parking lots, an empty McDonald's, a gas station, a strip mall, and a few scattered retail establishments, all car-oriented and set back from the street.

The sprawling entry point, called Kelley's Corner, is a short distance from the public school campus, and has been a source of embarrassment for townspeople for more than two decades. In a community outreach project organized with the Town Planning office in 2009, we asked residents what they wanted to see there instead. To encourage the widest response, we made relatively new computer simulation tools as well as old-fashioned art and craft supplies available for free. The citizens were then invited, literally, to redraw the map. We were surprised by how many people took an interest in this project. Several hundred residents participated in open meetings and more than fifty took the time and made the effort to create and submit an original plan (Sussman 2010). Not surprisingly, we learned that locals wanted Kelley's Corner to be walkable and family-oriented, with stores at the street, featuring commercial uses (like ice cream shops) that gave people reasons to linger. We learned that residents wanted focused activity specifically at the intersection that would mark it as a significant arrival point (Hollander *et al.* 2010). At the end of

Figure 5.6 A suburban intersection 40 km (25 miles) northwest of Boston, Kelly's Corner in Acton, Massachusetts, just off the highway to the city (Source: Ann Sussman).

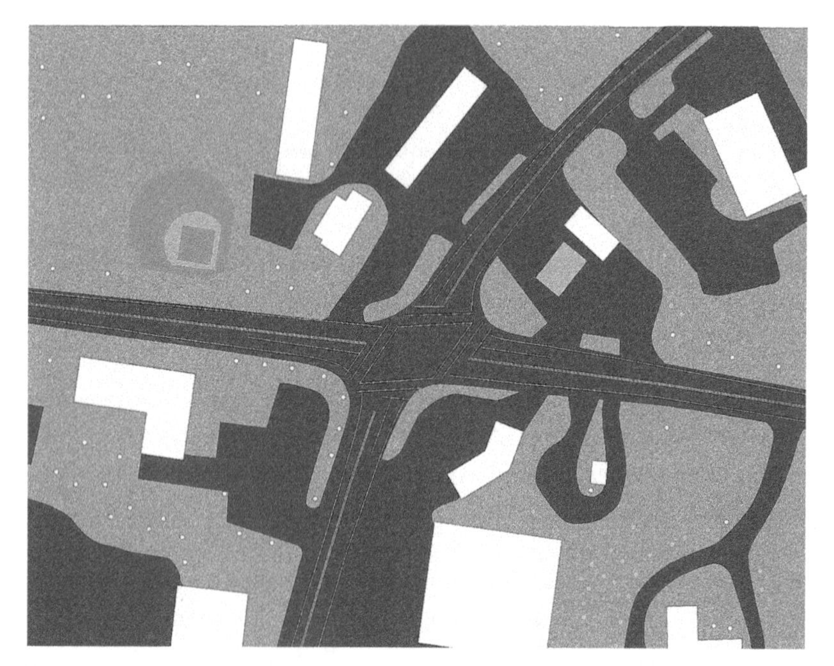

Figure 5.7 An aerial diagram of Kelly's Corner shows its scattered building layout where planning has accommodated parking lots and cars over pedestrian needs (Source: Justin Hollander and Amanda Garfield).

the exercise, we displayed all the citizen designs at a town forum and asked residents to vote for their preferred vision (see Figure 5.8). The citizens overwhelmingly preferred this one:

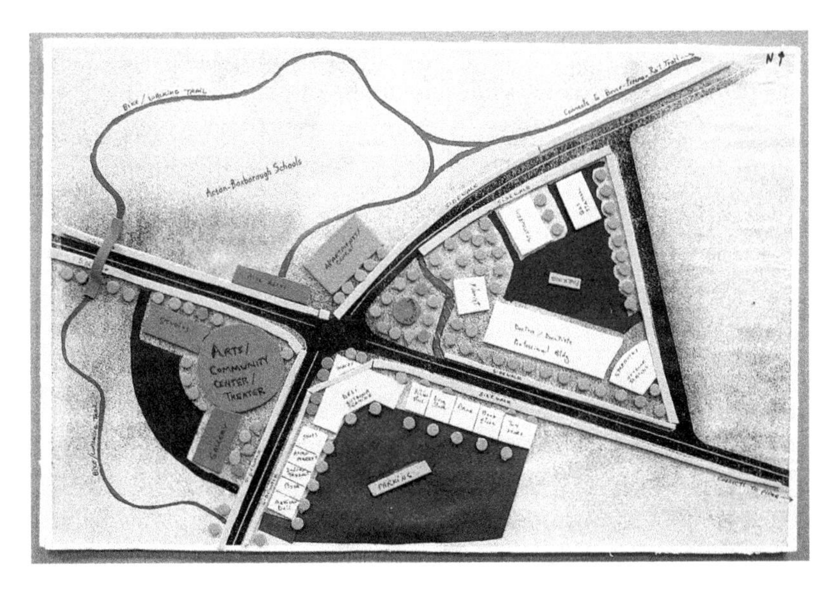

Figure 5.8 The winning Kelly's Corner plan was designed by residents Janice Ward and Mark Buxbaum into an arrival point and place of pride by focusing on building alignment and assigning a clear hierarchy that suggests an uplifting story (Source: Ann Sussman).

Significantly, not only does the winning scheme have the commercial and social spaces reflecting local wishes, but it also gives residents something more, something desired but not always clearly articulated—a sense of positive narrative. The plan that won is arranged hierarchically with the most important building (a large blue circle representing a community center with an arts and theatre component) dominant on one side of the central intersection and a fountain surrounded by trees on the other. It builds to an arrival point. The townspeople, it turned out, preferred not just more walkable retail or wider sidewalks, but something harder to specify—pride of place, a positive town identity, a winning story. The preferred scheme told anyone passing through Kelley's Corner that culture, education, and the arts mattered to people here, all placed in a thoughtfully arranged green setting. It would be hard to miss the way the narrative sequencing in the plan said that. This case study shows how important narrative is in giving a sense of importance to a site without one. People crave meaning in their built environments and, when given the option, innately select for it.

The Need for Narrative Expressed Poetically

The enduring need of our human species to make sense of our place in the world, and develop a story that secures it, is elegantly expressed in the poem, 'Things' by Lisel Mueller (b. 1924), a Pulitzer Prize-winning American poet. In the poem, reprinted below, she shows how we appropriate words that apply to our bodies to objects that surround us, and in so doing enable the creation of stories to help us frame our experience.

Things
by Lisel Mueller

What happened is, we grew lonely
living among the things,
so we gave the clock a face,
the chair a back,
the table four stout legs
which will never suffer fatigue.

We fitted our shoes with tongues
as smooth as our own
and hung tongues inside bells
so we could listen
to their emotional language,
and because we loved graceful profiles
the pitcher received a lip,
the bottle a long, slender neck.

Even what was beyond us
was recast in our image;
we gave the country a heart,
the storm an eye,
the cave a mouth
so we could pass into safety.[2]

Much as we anthropomorphize ‘things’ in our world, to make sense of it, we similarly look for our buildings and our urban places to reflect us and satisfy needs including our singular and voracious narrative appetite.

Exercise for Chapter 5: Storytelling is Key

- Select a favorite building or plan or monument and interpret it in terms of narrative sequencing. It may be useful to review a religious building this way.
- Choose a favorite building elevation; does it have a compelling heirarchy or suggest a narrative sequence?
- A 2013 *New York Times* article highlighting the popularity of eating smoked turkey legs in Disney theme parks, concluded: "Boiled down, Disney parks are about selling memories," and "story-telling" (Barnes 2013). Describe how and why this attribute is critical for theme parks and successful place-making in the built-environment generally.
- In the twenty-line poem 'Things' by Lisel Mueller, how many lines reference the face or human head? Is any body part named more often? Why might this be the case, particularly, given the research reveiwed in Chapter 3?

Notes

1 Its elements include: "1) the amygdalo-hippocampal system, responsible for initial encoding of episodic and autobiographical memories, 2) the left peri-Sylvian region, where language is formulated, and 3) the frontal cortices and their subcortical connections, where individuals and entities are organized into real and fictional temporal narrative frames" (Young and Saver 2001: 188).

2 'Things' by Lisel Mueller Reprinted by permission of Louisiana State University Press from *Alive Together* by Lisel Mueller. Copyright © 1996 by Lisel Mueller.

6

Nature is our Context: Biophilia and Biophilic Design

> ...with each new phase of synthesis to emerge from biological inquiry, the humanities will expand their reach and capability.
>
> E. O. Wilson (1984: 55)

The past chapters singled out several subconscious tendencies that govern our response to the built environment. Humans are thigmotactic, a 'wall-hugging' species. We are innately self-protective and tend to avoid the centers of places. We are a social species, visually attuned to take in other people, and specifically interpret their faces quickly. We tend to favor symmetrical shapes, curving forms, and visual complexity. We also love a good story to such an extent that we can appreciate one silently in the sequential flow of a building plan or garden design and conversely, get upset by a place that seems without one.

This chapter focuses on something different: our context, specifically the natural world we evolved in and how we love looking at that too (see Figure 6.1). It also reviews recent literature that reveals our inherent need to be connected to our context to nurture our mental, physical and spiritual health, and well-being.

We have an innate "tendency to focus on life and lifelike processes," biologist E. O. Wilson explains in his 1984 book *Biophilia* (p. 1). Wilson, now a Harvard professor emeritus, defines biophilia as the "urge to affiliate with other forms of

Figure 6.1 Back-yard arbor in Acton, Massachusetts; a predisposition for enjoying natural scenes is in our genome. Given the means, we embellish the view (Source: Ann Sussman).

life..." (p. 1). Our evolutionary past resonates daily with how we respond to our present environment, he persuasively argues. Evolving in the African and later European and Asian savanna, places with grassy plains with scattered trees, our hunter-gatherer ancestors (estimated to have lived from 1.8 million to 10,000 years ago) were consistently in contact with the natural world. The complexity of nature is the matrix where we, our genes, and our brains, came into their humanness. At the end of the day, we cannot be healthy, think well, and flourish by abandoning or ignoring our primal context.

"We stay alert and alive in the vanished forests of the world," Wilson wrote (1984: 101). Significantly, given the possibility of living anywhere, people still, so many thousands of years later, "gravitate statistically" toward a savanna-like environment, he noted (see Figure 6.2). The habitat preference suggests how our

Figure 6.2 The acacia tree common in a savanna is thought to have been one common element in our primal vista (Author: Neelix, Source: Wikimedia Commons).

distant past lives on in our present. Wilson, later writing with co-author Stephen Kellert, isolates three features that impact modern views of an ideal setting; tellingly, all derive from our ancestral vista:

> . . . people want to be on a height looking down, they prefer open savanna-like terrain with scattered trees and copses, and they want to be near a body of water, such as a river or lake, even if all these elements are purely aesthetic and not functional. They will pay enormous prices to have this view.
>
> (Kellert and Wilson 1993: 23)

The case Wilson makes is that we can never leave our evolutionary past behind and that we hurt our species' prospects if we try to. What we can do is work to understand, celebrate, and promote our connections and dependence on nature. This is the tack taken by practitioners of biophilic design. "(T)he human mind and body evolved in a sensorial rich world," which remains significant for our "health, productivity, emotional, intellectual and even spiritual well-being," wrote Stephen Kellert, Judith Heerwagen, and Martin Mador (2008) in their introduction to *Biophilic Design: The Theory, Science and Practice of Bringing Buildings to Life* (p. vii). Advocates of the biophilic approach, such as Kellert, professor emeritus at Yale's School of Forestry, and colleagues, critique most modern building practices for promoting a sense of 'placelessness' in the built world: too many new buildings contribute to the visual impoverishment of our cities and towns; too many urban centers heedlessly replicate the concrete canyons of Boston City Hall environs or worse (see Figure 6.3 and Chapter 2). While obviously meeting basic habitat requirements, these structures betray us in the end. The consequences of their construction include "sensory deprivation, where monotony, artificiality and the widespread dulling of the human senses are the norm rather than the exception" Kellert (2012) wrote in *Birthright: People and Nature in the Modern World*, as though channeling Jane Jacobs a half century ago (p. 161).

Aesthetics matter, biophilic designers say. Indeed, they criticize the growing green-building movement, which was galvanized by the formation of the non-profit USGBC in 1993, for not going far enough and mentioning aesthetics. The sustainability movement's work to encourage resource conservation is critical, Kellert writes. Green building has changed the conversation about construction in the United States in the past two decades, but the movement needs to embrace the significance of personal contact with nature, and of building within a "culturally and ecologically relevant context" (Kellert 2012: viii). Without considering how people relate to buildings and appreciate certain forms over others, green building hobbles its own

Figure 6.3 Boston City Hall, at far left, and its immediate surroundings. Repetitive machine-like elevations disconnect us from nature; the buildings do not resemble the lifelike forms and landscapes we evolved within (Source: Ann Sussman).

chances of future success. The green movement fails to achieve its goal of sustainability, Kellert explains, because it falls short of nurturing the physical and mental benefits that create emotional attachment to place in the first place, and then motivates people to care for their constructions and retain them over the long term (Kellert 2012: 162).

What makes the biophilic advocates' argument difficult to dismiss is their consideration and dedication to science, including both new research and older evolutionary theory. Acknowledging that nature can also be harsh and scary, (people across cultures have a fear of snakes and spiders, for instance) biophilic designers explain that setting out to control the environment is a hallmark of many species, not only human beings. "The urge to master nature is a normal and adaptive tendency,"

Kellert wrote in *Birthright*. "Keystone species," including elephants, beavers, sea otters, termites, and *H. sapiens*, are adept at "reshaping their world" (Kellert 2012: 81). However, "no other creature has so mastered and controlled its environment as have modern humans, arguably to an excessive and dysfunctional degree." It is imperative for all life that our constructions align more carefully and better recognize our place within the natural world.

In support of the biophilic approach, its advocates have mined the historical record. They note how two thousand years ago, Chinese Taoists noticed gardens carried health benefits (Wilson 2006: 326). In Medieval Europe, monks created elaborate monastery gardens, in part understanding that these soothed the sick (Ulrich 1984: 2). In England, famed nurse Florence Nightingale wrote in her book *Notes on Nursing*, published in 1860, that "variety of form and brilliancy of color in the objects presented to patients are an actual means of recovery" (Nightingale 1860: 59, as quoted in Wilson 2006: 326). These early observers laid the foundation for the biophilic approach.

Now modern research is demonstrating the validity of some of their old insights. Studies show that experiencing nature and its visual complexity, even for a few minutes, even represented in a painting or photograph, reduces stress and has other benefits. In one research study, for instance, anxious patients in a dental clinic were measurably less nervous on days when "a large nature mural" showing a leafy scene decorated the waiting room than on days when the wall was blank (Heerwagen 1990). Another frequently cited research paper by Roger Ulrich, of Texas A & M University, carefully monitored the recovery rates of patients after gall-bladder surgery. It found that those patients with hospital windows facing trees "had shorter hospital stays and suffered fewer minor post-surgical complications," than those whose windows faced a brick wall (Ulrich 2002: 7). The lucky patients with green views also requested and took less pain medicine. This type of research obviously has huge implications for the billion-dollar health-care industry and has led to the growing adoption of 'evidence-based design' in health-care facilities.

More recent studies show exposure to outdoors as important for child health and development, (Wilson 2006) and link the lack of it to a rise in attention deficit hyperactivity disorder (ADHD) in American children. Researchers also report adopting the biophilic approach in office interior design can generate a quantifiable boost in worker productivity.[1]

Kellert and other biophilia advocates believe the way forward, elaborating somewhat on Jacobs' insight that man was "a part of nature," is for humans to seek "reconciliation if not harmonization with nature" (Kellert 2012: 5). To promote the biophilic approach, Kellert outlines six key elements and more than 70 attributes, suggesting a diversity of feasible approaches and styles in the built environment. In abbreviated form, these are outlined in Table 6.1.

This book promotes the biophilic approach and focuses on

Table 6.1 Key Elements of Biophilic Design

1	Environmental features, such as plants, water, sunlight;
2	Natural shapes and forms, including botanical and animal motifs, shells, and spiral forms;
3	Natural processes and patterns; similar forms at different patterns and scales; sensory variability;
4	Light and space; natural light; inside-out spaces;
5	Place-based relationships; geographical connection to place; historical connection made manifest;
6	Evolved human relationships to nature; prospect and refuge, order, and complexity.

Source: Kellert 2012: 171–2

our subconscious human responses to places to most effectively do so. Essentially we take an inside-out approach to the problem of solving the riddle of how to best design for humanity. We argue that it is best to first look at how people are built—not only mechanically but also mentally, subconsciously, and then design or plan for these requirements and tendencies. In teasing apart the evolutionary scrim people look through, the intent is to encourage many more creative approaches to the task, and not proscribe one specific architectural or planning style. With this end in mind, each of the preceding chapters can be considered a rule-of-thumb for framing human behavior in the built environment. These are reiterated below, in the order they appeared in the book:

- Edges Matter: This principle, discussed in Chapter 2, describes how as pedestrians we are a 'wall-hugging' species. The more designers are aware of thigmotaxis as a billion-year-old biological trait, the better they will understand why well-defined corridor streets encourage our walking and the imperative of creating them in suburban and urban places;
- Patterns Matter: This principle, outlined in Chapter 3, reminds us that the human mind prioritizes vision. We evolved in a world of visual complexity and relish visual stimulation, not sameness nor blankness. We are also biologically designed to process, emotionally engage with, and remember facial pattern over other forms. The template for the face is with us from infancy;
- Shapes Carry Weight: This principle, discussed in Chapter 4, recognizes that humans are programmed to prefer certain forms over others; we carry innate biological biases toward bilateral symmetric shapes, and for curved versus straight, or jagged lines or forms; and
- Storytelling is Key: This principle, reviewed in Chapter 5, most identifies us as human; our narrative capacity, a consequence of our species' unique neural circuitry, helps us

engage with others, with places, with a shared past and enables the creation of identity. Most popular designed places engage this aspect of human uniqueness in some manner.

Taken together, this book's principles loosely prescribe a way of thinking about human behavior with the hopes of building a foundation for the creation of better designs and more compelling places. "The broader one's understanding of the human experience, the better design we will have," Steve Jobs, Apple CEO once said describing the essential idea at the base of his company's successful computer products (Thomas 2011: 71). In this book we have outlined the salient, mostly hidden, aspects of human experience, the ones we believe are significant for planning and architecture, with the same intention—it will lead to better design. What the research discussed here says, incontrovertibly we think, is that human beings evolved to look at nature and make deep connections with other human beings. Man-made environments that reflect these facts in various ways (and there are many ways to go about it, as some of the book's case studies illustrate) will be more sustainable. They will be places that more likely enhance our lives, will more likely be cared about, and, in the end, more likely to last.

Implications for Practice

Many considerations govern the details of designing and planning in the built environment. We are not suggesting an abandonment of other design imperatives, but instead a need to integrate those requirements with what research shows us about who we are and how we came to be. Today's practitioner who is shaping buildings, streets, and neighborhoods can no longer rely on influences that are simply tasteful or utilitarian. An opportunity is available to "design for people" in a way that noted Danish architect Jan Gehl promotes. By grounding our designs in our evolution, we believe there is the possibility to create more enduring, sustainable places that people will enjoy inhabiting.

Implications for Policy

One of the most powerful influences on the built environment is the (primarily) local regulatory system that controls the use of land, building form and bulk, and parking (Platt 1996). The principles outlined in this book offer critical lessons for adjusting policy at the local, state, and federal levels. Most decisions about how the built environment is designed and shaped are made at the local level and that is where the ideas of this book can make an impact. Zoning, subdivision control, and design review processes provide local elected or appointed officials the chance to guide and shape how development happens, subject to state and federal guidelines. In much of the U.S., that local authority is strong and cities have great latitude to control form, function, bulk, and use. Local officials may wish to consider these principles and look for ways to update their rules and regulations.

State and federal agencies typically fund much of the transportation infrastructure in the U.S. and, globally, these transportation officials ought to understand that their policy decisions will shape how well places work for people. For urban planners, community-wide visioning exercises may profit by beginning with a look at the principles elaborated here. That way clients are more likely to believe design professionals truly have their interests at heart. Before the first public meeting, before goals are identified for community growth or change, it may prove useful for planners to look to these principles as a foundation for understanding how people experience places. From there, it is then possible to explore goals, values, and the development of a plan for the future.

Implications for Theory

Theory helps us to understand the relationship between concepts, why things happen, how and by whom. Here, this book's principles may have the potential for the greatest impact. When designers and planners consider the question of what makes a place great or how to rate a public design, we believe they now have a better starting point. This ethic allows for the ultimate judge of success to be grounded in answering the question: how well does this place meet the needs of an evolved, bipedal mammal? How will it be received by people never met?

Figure 6.4 The Great Workroom, SC Johnson, a family-owned company in Racine, Wisconsin by architect Frank Lloyd Wright, 1936, has been called "the most beautiful office space in America." Wright called the columns, "dendriform," or tree-shaped; some see them as lily pads. They are 5.6 meters (18.5 feet) in diameter at the top and 23 centimeters (about 9 inches) in diameter at the base (Photo credit: SC Johnson).

Figure 6.5 Columns in thirteenth-century Chartres Cathedral interior, Chartres, France. The space emotionally engages us no matter our background or creed; it reads as a sacred forest (Source: Rama, 'Own work,' Wikimedia Commons, August 11, 2009).

Figure 6.6 Five- to seven-hundred-year-old Redwoods in Humboldt Redwoods State Park, Northern California. It has been said that nature provides the template for our most elegant architecture (Source: Jason Sturner (www.flickr.com/photos/50352333@N06/4644487863/), Wikimedia Commons, June 28, 2008).

Exercise for Chapter 6: Nature is our Context

- In the photographs on preceding pages (see Figures 6.4–6.6), analyze how the natural world provided a template for each design; also describe aspects of the design that specifically address human needs and are not found in nature.
- Select a favorite building or urban plan and analyze it in terms of the four principles articulated above that aim to describe human experience in the built environment: Edges Matter, Patterns Matter, Shapes Carry Weight, and Storytelling is Key. How is this approach useful? Could it be improved? What does it overlook?

Notes

1 In one study of employees at the Herman Miller office furniture company, it was found that adding interior vegetation, natural light, and outdoor seating to an office measurably increased worker productivity and well-being. (greenplantsforgreenbuildings.org/green-plant-benefits/)

Appendix

More on the Morphology and Function of the Human Brain

The human brain makes people outliers in the animal kingdom. It is also a mystery to us, like an iceberg, mostly hidden, as Freud said. There are several models for thinking about the brain. A typical diagram, which masks the complexity, looks like Figure A1.

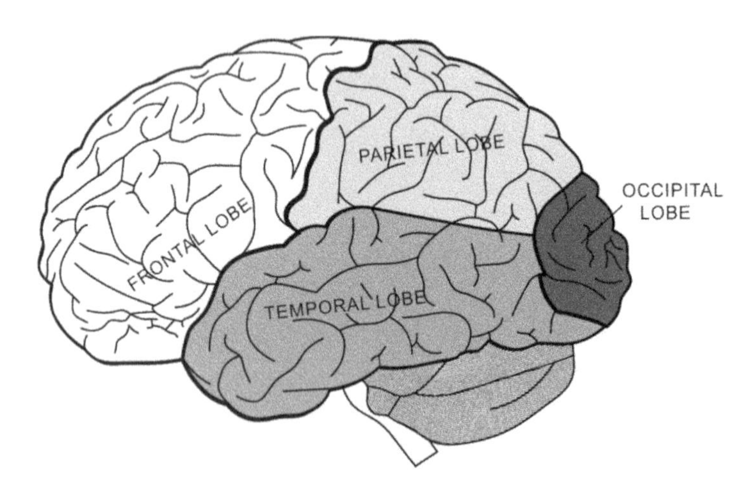

Figure A.1 The brain is often described as having four lobes, each with fairly specific functions: the frontal lobe is responsible for conscious thought and mood; the parietal lobe plays a role integrating sensory information from the various senses and manipulating objects; the occipital lobe is concerned with visual processing; and the temporal lobe is involved in retaining visual memories, understanding language processing, and the processing of complex visual stimuli such as faces (Source: Trey Kirk).

This shows the cerebral cortex, the outer-most folded layer of the brain, which enables complex thinking, with various 'lobes' that take their names from the parts of the skull covering them. A well-known feature of the brain is its size, particular its relative size. The human brain is larger than any other mammalian brain when body mass is taken into account. (We have a smaller brain than do elephant or whales, for instance, but they are larger mammals). The encephalization quotient, (EQ) which takes relative size into account, is shown in the Figure A.2. The gorilla, a fellow primate and mammal, which can be as large as a person or larger, has a brain one-third the size of our own (Fonseca-Azevedo and Herculano-Houzel 2012) (see Table A.1).

But new research shows size is not everything—it is the specialized circuitry that gives us our humanness, as Michael S. Gazzaniga and co-authors describe in *Cognitive Neuroscience: The Biology of the Mind*, 3rd ed (2009):

Figure A.2 Diagram of the encephalization quotient (EQ) in various animals. The human brain is an outlier at three times the size of its nearest relation on the evolutionary tree (the chimpanzee) when factoring in body size (Source: Trey Kirk).

> Complex capacities like language and social behavior are not constructs that arise out of our brain simply because it is bigger than a chimpanzee's brain. Instead, these capacities reflect specialized devices that natural selection has built into our brains.
>
> (p. 664)

Combined with its special circuitry, another uniqueness of the primate brain is that it contains more neurons, or nerve cells, than a similarly sized rodent brain—as much as seven times more according to one analysis (Fonseca-Azevedo and Herculano-Houzel 2012). And a further distinction of the human brain is that it contains appreciably more neurons than that of any other primate or creature on earth. The adult male human brain has an estimated 86 billion nerve cells. These in turn are capable of making trillions of synapses, chemical, or electrical connections to other cells. And, connections are everything, as they say. However, we pay a price for all this activity. The human brain, about 2% of body mass, uses more energy than any other organ, accounting for 20 to 25% of the body's energy intake (see Table A.1). The brain needs the energy to both fire and maintain its horde of nerve cells. Our intelligence comes at a cost—no wonder we avoid stairs!

The Significance of Cooking

Table A.1 shows how we are like other animals—and not. The enormous brain size, three times the size of our nearest relative on the evolutionary tree, in some senses also may contribute to making us feel more different from other creatures than our bodies are. Intriguingly, the energy appetite of the brain has promoted evolutionary theories about how the human brain came to be and whether it could get larger in the future. "It may have been a change from a raw diet to a cooked diet that

Table A.1 Brain Size and Energy Consumption Table

Animal	Encephalization Quotient[1]	Brain Energy Consumption[2]
Human	7.4–7.8	20–25%
Chimpanzee	2.2–2.5	8-10%
Gorilla	1.5–1.8	
Whale	1.8	
African elephant	1.3	
Dog	1.2	
Squirrel	1.1	3–5%
Horse	0.9	
Mouse	0.5	
Rabbit	0.4	
Rat	0.4	

This table shows the relationship between brain sizes adjusted for body mass and energy consumption. The encephalization quotient (EQ) is the ratio of an organism's brain size to its body mass. A higher EQ generally indicates greater intelligence. Brain energy consumption is the amount of the body's total resting energy that the brain uses to function. A higher percentage means that more energy is devoted to brain functioning (Roth and Dicke 2005; Snodgrass *et al.* 2009).[1,2]

Source: Devin Merullo

afforded its remarkable number of neurons," said Karina Fonseca-Azevedo and Suzanna Herculano-Houzel in a 2012 issue of the *Proceedings of the National Academy of Sciences*. Since cooking enormously increases the energy yield of foods and the speed with which we can consume them, this theory holds that using fire to make food was the crucial step in promoting "the near doubling of numbers of brain neurons" our ancestors experienced. It also can help explain why other creatures have not overtaken us in brain power: they have not learned to master fire and cook.[3] The epic transformation is estimated to have occurred between *Homo erectus* (c. 1.8 million years ago) and *H. sapiens* (c. 200,000 years ago).[4] The elephant, for example, has the largest size brain for a land mammal, weighing in it 4 kg or more, (8 lbs) and has to spend most of its time, some sixteen to eighteen hours each day, feeding to maintain brain and body health (Sea World 2014). Humans standing upright have freed up their appendages to put together ingredients, heat them, and effectively pre-digest them, outsourcing part of the digestive process outside the body. This is one reason, scientists believe, humans do not have the enormous digestive systems other large animals such as elephants and cows do and do not spend hours ruminating, in the digestive sense, either.

Another way to describe the brain is to look at it in section. And a typical section looks like Figure A.3.

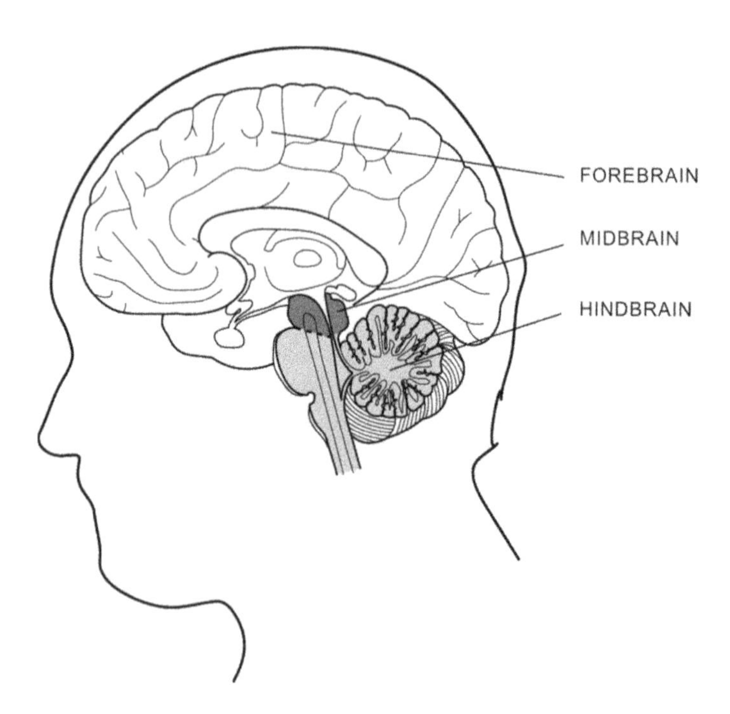

Figure A.3 Diagram of the human brain in section shows the forebrain, midbrain, and hindbrain (Source: Trey Kirk).

Figure A.3 shows the forebrain, midbrain, and hindbrain. These basic elements we share with other animals, although their forebrains are smaller. Starting from the top, the forebrain is responsible for thinking and processing sensory input and makes up the left and right cerebral hemispheres. It is covered by the highly ridged pinkish-gray cerebral cortex. Nested below the cortex are five diverse structures that take part in higher level thinking including the hippocampus and the amygdala, responsible for forming memories and emotional regulation respectively.

The lobes of the cortex include the frontal lobes that are involved in planning future action; the parietal lobes further back on the head, involved with body awareness and manipulating objects; the occipital lobe, concerned with processing visual information; and the temporal lobes at the side, also involved with memory recall and interpreting visual information, including facial recognition. The point is each region of the brain has pathways to expertly handle specific tasks. The midbrain, the smallest region deep within the brain, controls eye movement. The hindbrain, which extends up from the spinal cord, is responsible for activities considered lower-order and more automatic, including breathing, heart rate, blood pressure, and maintaining balance and equilibrium.

Other models of the brain underscore the evolutionary timeline we share with our predecessors on our evolutionary tree. For example, the Triune Brain model, developed by Dr Paul MacLean in the 1960s, organizes the brain as a compilation of three brains, the oldest at bottom, the reptilian brain, or brain stem, followed by the mammalian brain, or limbic brain, and topped off by the human brain, or neo-cortex (see Figures A.4 and A.5).

In this model, the reptilian brain controls primitive behaviors such as survival and aggression, the paleomammalian complex is responsible for social behavior and simple emotions, and the neomammalian brain enables complex thought. Although an

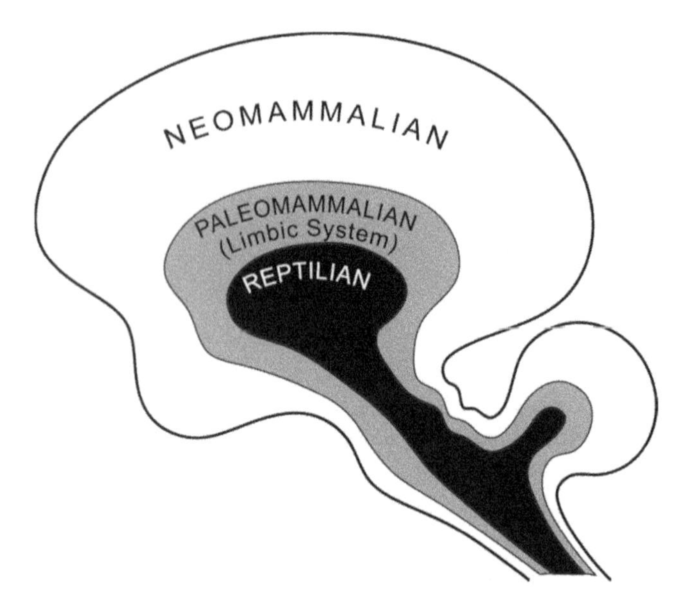

Figure A.4 A diagram of the 'Triune' Brain Model, developed by the physician Dr Paul MacLean, underscores the evolutionary past we share with other creatures on earth and models the brain as a series of sequential additions. (Source: Trey Kirk).

oversimplification, this model can be used as a loose analogy to recognize what various parts of the brain do. While we do not literally have reptile and early mammal brains connected to a modern human brain, these three different regions of our brains play roles in regulating a range of our behaviors, from the instinctive to more measured.

The point is not only that our brain organization mimics that of other creatures that we may not have a particular affinity to, such as the snake and monkey shown in Figure A.5, but we also share some behaviors with these creatures. Since evolution is conservative, traits that work reappear. We might say that our reptilian brain (snake-like) is at work when a sudden pang of hunger strikes us around 6 o'clock in the evening and all we can think of is food and how to get it fast. But our protomammalian

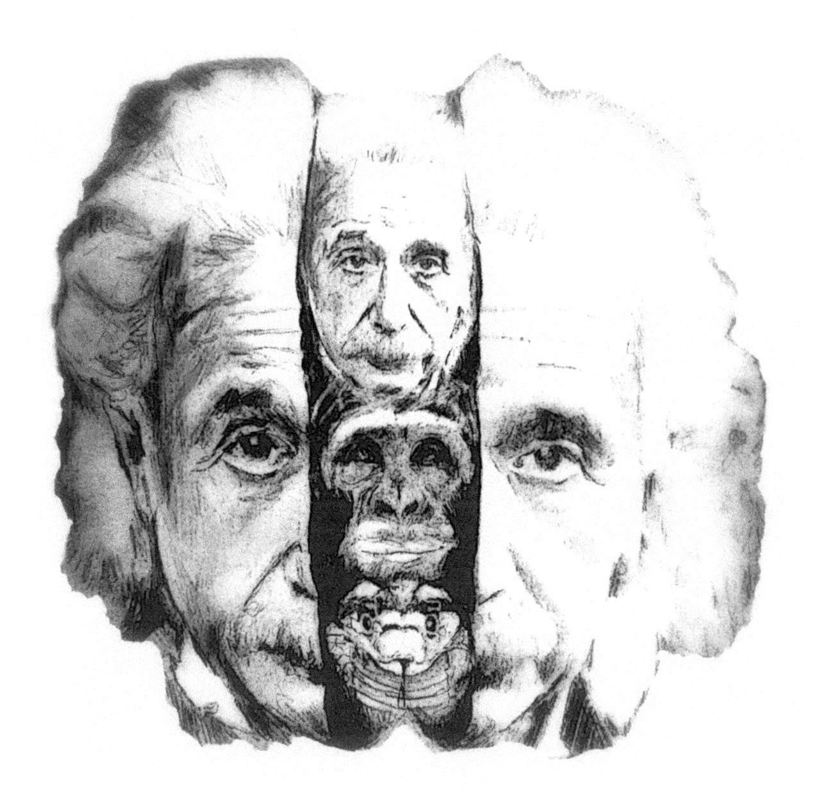

Figure A.5 The 'Triune' Model as a schematic portrait of Albert Einstein: within us are elements of animals that came before as rendered by artist Trey Kirk.

brain (monkey-like) kicks in when we stop to consider our spouse, children, or other associates and what their preferences for dinner might be. It is our neomammalian brain (the human) that enables to weigh all possible options and make the executive decision on what the best choice may be. Clearly the brain is complex. And, in terms of our sensory inputs, the brain does not prioritize them equally, as discussed in Chapter 3. The human brain prioritizes vision, and is always alive to the possibility of new arrangements in our field of view; all this, of course, a product of our singular evolution.

Notes

1 Roth, G., and Dicke, U. 2005. Evolution of the brain and intelligence. *TRENDS in Cognitive Sciences*, 9(5), 250–7.
2 Snodgrass, J. J., Leonard, W. R., and Robertson, M. L. 2009. The energetics of encephalization in early hominids. In *The Evolution of Hominin Diets*, pp. 15–29. Springer Netherlands.
3 The high-energy requirement of our brains may indicate we have reached a physical limit: the end of feasible brain expansion in the future. (This precludes the idea that we will soon have brains with computer chips in them.) We can't get smarter because human brains would consume too much energy, said Simon Laughlin, professor of neurobiology, at Cambridge University, who told *The Sunday Times*: "We have demonstrated that brains must consume energy to function and that these requirements are sufficiently demanding to limit our performance and determine design" (July 31, 2011). We are, in other words, maxed out (Leake 2011: 1, 5).
4 There is debate in the literature about when our human ancestors first got control of fire, ranging from 1.8 million to 400,000 years ago.

Acknowledgments

This book could not have happened without the consistent efforts and energetic input of many people including research assistants at Tufts: Elza Lambergs, Caroline Geiling, Annie Levine, Margaret Wiryaman, Jingyu Tu, Gabriel Holbrow, and Acton high school senior, Julia Call. Devin Merullo, biology graduate student at University of Wisconsin, performed a virtuoso task translating research on the brain, neuroscience, and psychology into vernacular English. Harvard architecture graduate student Nora Shull and recent grad Trey Kirk seemingly effortlessly produced the many black and white diagrams and figure-ground drawings that accompany the text and bring it to life. We especially call out Nora's 'Thatcherized Face' and Trey's hand-drawn 'Einstein.' We have it on good authority that anyone seeing those faces will never forget them. Special thanks for administrative support go to Maria Nicolau, Doug Kwartler, and Ann Urosevich of Tufts for all their help and good humor along the way.

For the many photographic contributions, we thank architect Garry Harley for the professional photographs of Italy and France, Deniz Gecim for the images from Turkey, and Celia Kent for the shot of the English village. We would like to thank Professor Stephan Chalup, Associate Professor in Computer Science at the University of Newcastle, Australia, for sharing his photographs of happy-looking houses on the other side of the world. We also thank Rodrigo Cardenas, post-doc at Penn State University and Lauren Julius Harris, Professor of Psychology at Michigan State University for sharing their work and, in particular, Dr Cardenas' unique photographs of symmetrical and asymmetrical faces and pattern in indigenous craft. In Brooklyn, NY, we thank Eve Sussman and Simon Lee for providing lodging in the city, which gave us time to explore and photograph Jane Jacobs' remarkable old West Village neighborhood. At the Bradford Mill in West Concord, we thank Sam of Palm Press Atelier for making photoscanning a pleasure. In Melbourne, Australia, we want to particularly thank Simona Castricum of ARM Architecture for providing us with a photo of the new Portrait Building. In Philadelphia, we thank Hamil Pearsall of Temple University for helping us secure photos of Society Hill. At *Architect, the Magazine of the American Institute of Architects*, we thank editor Ned Cramer and Senior Graphic Designer Alice Ashe for providing a photo of the Society Hill neighborhood. We also thank the George Cserna/Avery Architectural and Fine Arts Library, Columbia University rights to use the photo in this book. Also, many thanks to Erica Bossier and LSU Press for permission to reprint the poem, 'Things.'

In preparing this text, we thank Duke University Professor Adrian Bejan for answering our several emails about the golden rectangle and the 'constructal law,' and Professor Holly Taylor of Tufts. We also thank Dr W. Jake Jacobs, Psychology Professor at the University of Arizona for sharing his work on thigmotaxis and responding to our early emails about this project. We thank Dr Lynn Nadel, Professor of Psychology and Cognitive Science, also at the University of Arizona in Tucson, for speaking with us about thigmotaxis and its function in mammals of prey. We thank Professor Rusty Gage of the Salk Institute for Biological Studies, past president of Academy of Neuroscience for Architecture (ANFA) for a lengthy phone interview on this project in its incipient stages last spring. We also thank neuroscientist Dr Eric Kandel for helping us make the connection to Dr Gage. We are also indebted to Dr Kandel for his 2012 book, *The Age of Insight: The Quest to Understand the Unconscious in Art, Mind and Brain from Vienna 1900 to the Present*, without which the present text would not be possible.

We also must thank the Cloud Foundation and ArtScience Prize in Boston (www.artscienceprize.org/boston) for sparking creativity in young people and not-so-young facilitators. Working with students at the Cloud Foundation in 2010 provided the "seed idea" for this work, a new kind of book on the brain, evolution, and architecture.

We thank our editors at Routledge, Wendy Fuller, Emma Gadsen, and Rebecca Hogg for their support and for helping find a workable title for our ideas. We also thank Susan Schulman for her early interest.

Closer to home we thank our first and extremely loyal readers, Janice Ward, Jane Ross, and Celia Kent. Their interest and comments turned out to be invaluable. Any writer could use this group. We thank the Nashoba Brook Bakery in West Concord for providing not only great fare, but with their brook-adjacent windows, a sustaining vista, and great place to write. Lastly, but most significantly, we thank our families, specifically in Grafton, Pam, Rose, and Sam, and in Concord, Chris, Ben, and Tom for their endless (well, almost) patience. "You are the sauce for my spaghetti."

References

Aldersey-Williams, Hugh. 2013. *Anatomies: A Cultural History of the Human Body*. New York: W.W. Norton and Company.

Alexander, Christopher, Sara Ishikawa, and Murray Silverstein. 1977. *A Pattern Language: Towns, Buildings, Construction*. New York: Oxford University Press.

Appleton, Jay. 1975. *The Experience of Landscape*. London: John Wiley and Sons.

Arnheim, Rudolf. 1969. *Visual Thinking*. University of California Press.

Bacon, Edmund N. 1974. *Design of Cities*. Thames and Hudson.

Banich, Marie T., and Rebecca Jean Compton. 2010. *Cognitive Neuroscience*. CengageBrain.com.

Bar, Moshe, and Maital Neta. 2007. Visual elements of subjective preference modulate amygdala activation. *Neuropsychologia*. 45 (10): 2191-200.

Barnes, Brooks. December 27, 2013. "Turkey Legs Conquer Land of Mouse Ears." *New York Times*. Available at: www.nytimes.com/2013/12/28/business/media/disneys-newest-star-turkey-leg-wins-hearts-and-stomachs.html?_r=02 (accessed May 26, 2013).

BBC News. November 23, 2004. "'Virgin Mary' toast fetches $28,000." Available at: http://news.bbc.co.uk/2/hi/4034787.stm (accessed May 26, 2013).

Bloom, Nicholas Dagen. 2001. *Suburban Alchemy: 1960s New Towns and the Transformation of the American Dream*. Urban Life and Urban Landscape Series. Columbus: Ohio State University Press.

Boston Globe. January 13, 2013. "Ada Louise Huxtable was the Architecture Critic Who Loved City Hall." Available at: www.bostonglobe.com/editorial/2013/01/13/ada-louise-huxtable-was-architecture-critic-who-loved-city-hall/vv9z87NDhG5cQJ9JVTAkHI/story.html (accessed March 25, 2014).

Chalup, Stephan K., Kenny Hong, and Michael J. Ostwald. 2010. Simulating pareidolia of faces for architectural image analysis. *Brain* 26 (91): 100.

Dahl, Christoph D., Nikos K. Logothetis, Heinrich H. Bülthoff, and Christian Wallraven. 2010. The Thatcher illusion in humans and monkeys. *Proceedings of the Royal Society of London. Series B: Biological Sciences* 277 (1696): 2973–81.

Darwin, Charles. 1859. *On the Origin of Species by Means of Natural Selection*. New York: Signet Classics, 2003.

Darwin, Charles. 1882. *The Descent of Man, and Selection in Relation to Sex*. New York: D. Appleton and Company.

De Botton, Alain. 2006. *The Architecture of Happiness*. New York: Pantheon.

Dixon, John Morris. January 1, 2014 "Philadelphia Resurgent." *Architect*. Available at: www.architectmagazine.com/multifamily/philadelphia-resurgent_o.aspx (accessed May 25, 2014).

Emrath, Paul. February 11 2009. How Long Buyers Remain In Their Homes. In: HousingEconomics.com. Available at www.nahb.org/generic.aspx?sectionID=734&genericContentID=110770&channelID=311 (accessed December 1, 2013).

Ewing, Reid, and Keith Bartholomew. 2013. *Pedestrian and Transit-oriented Design*. Urban Land Institute and American Planning Association.

Finnerty, John R., Kevin Pang, Pat Burton, Dave Paulson, and Mark Q. Martindale. 2004. Origins of Bilateral Symmetry: Hox and dpp expression in a sea anemone. *Science* 304 (5675): 1335–7.

Fonseca-Azevedo, Karina, and Herculano-Houzel, Suzana. 2012. Metabolic constraint imposes tradeoff between body size and number of brain neurons in human evolution. *Proceedings of the National Academy of Sciences of the United States of America*. 109; 45.

Gazzaniga, Michael S., Richard B. Ivry, and George R. Mangun. 2009. *Cognitive Neuroscience: The Biology of the Mind*. 3rd ed. New York: Norton.

Gehl, Jan. 2010. *Cities for People*. Washington, DC: Island Press.

Geidion, S. (n.d.) *First Gropius Lecture 1961*. Cambridge: Harvard University.

Goldfield, David R. 2007. *Encyclopedia of American Urban History*. Thousand Oaks: Sage Publications.

Gordon, Kate. 1909. *Esthetics*. H. Holt, New York USA.

Hazzard, Frank. Expert: Executing the Town Center Plan Essential to Columbia's Future. June 5, 2011. *Columbia Patch*. Available at http://columbia.patch.com/groups/politics-and-elections/p/expert-executing-the-town-center-plan-essential-to-coa55b1d2c0d

Heerwagen, Judith H. 1990. The Psychological Aspects of Windows and Window Design. Paper presented at proceedings of 21st annual conference of the Environmental Design Research Association. Oklahoma City: EDRA.

Hildebrand, Grant. 2008. *Biophilic Architectural Space*. Hoboken, NJ: John Wiley & Sons.

Hollander, Justin, Amanda Garfield, Yun Luo, Nina Birger, Kate Seldon, Andy Likuski, Becky Gallagher, Michelle Moon, and Pete Kane. 2010. Lessons from Open Neighborhood/Re-visioning Kelley's corner: An experiment in public participation. Report to the Town of Acton, MA.

Holt, Sandra. May 6, 2012. "Palmer Square to Celebrate 75th Anniversary." *Princeton Patch*. Available at: http://princeton.patch.com/groups/business-news/p/palmer-square-to-celebrate-75th-anniversary (accessed May 25, 2014).

Jacobs, Jane. 1961. *The Life and Death of Great American Cities*. Vintage.

Kahneman, Daniel. 2011. *Thinking, Fast and Slow*. Macmillan.

Kallai, Janos, Tamas Makany, Arpad Csatho, Kazmer Karadi, David Horvath, Beatrix Kovacs-Labadi, Robert Jarai, Lynn Nadel, and Jake W. Jacobs. 2007. Cognitive and affective aspects of thigmotaxis strategy in humans. *Behavioral Neuroscience* 121 (1): 21.

Kallai, Janos, Tamas Makany, Kazmer Karadi, and William J. Jacobs. 2005. Spatial orientation strategies in Morris-type virtual water task for humans. *Behavioral Brain Research* 159 (2): 187–96.

Kandel, Eric R. 2012. *The Age of Insight: The Quest to Understand the Unconscious in Art, Mind and Brain from Vienna 1900 to the Present*. 1st ed. New York: Random House.

Kanwisher, Nancy, Josh McDermott, and Marvin M. Chun. 1997. The Fusiform Face Area: A module in human extra striate cortex specialized for face perception. *The Journal of Neuroscience* 17 (11) (June 01): 4302–11.

Kastl, Albert J., and Irvin L. Child. 1968. Emotional meaning of four typographical variables. *Journal of Applied Psychology* 52 (6, Pt.1): 440–6.

Kellert, Stephen R. 2012. *Birthright: People and nature in the modern world*. New Haven, CT: Yale University Press.

Kellert, Stephen R., Judith Heerwagen, and Martin Mador. 2008. *Biophilic Design: The theory, science, and practice of bringing buildings to life*. Hoboken, NJ: John Wiley & Sons.

Kruh, David. 1999. *Always Something Doing: Boston's Infamous Scollay Square*. UPNE: Boston.

Lazzaro-Bruno, Claudia. 1977. The Villa Lante at Bagnaia: An allegory of art and nature. *The Art Bulletin* 59 (4): 553–60.

Leake, Jonathan. Brain power finally runs out of puff; That's our lot—humans can't get any smarter. *The Sunday Times* (London). July 31, 2011: 1, 5.

Leinberger, Christopher B. 2008. The Next Slum? *Atlantic Monthly*. Available at: www.theatlantic.com/magazine/archive/2008/03/the-next-slum/306653/

Lieberman, Daniel. 2013. *The Story of the Human Body: Evolution, health, and disease*. Harvard Museum of Natural History.

Little, Anthony C., and Benedict C. Jones. 2003. Evidence against perceptual bias views for symmetry preferences in human faces. *Proceedings of the Royal Society of London. Series B: Biological Sciences* 270 (1526): 1759–63.

Makin, Alexis D. J., Moon M. Wilton, Anna Pecchinenda, and Marco Bertamini. 2012. Symmetry perception and affective responses: A combined EEG/EMG study. *Neuropsychologia* 50 (14): 3250–61.

Marling, Karal Ann, and Centre canadien d'architecture. 1997. *Designing Disney's Theme Parks: The Architecture of Reassurance*. Montréal: Centre canadien d'architecture/Canadian Centre for Architecture.

McKone, Elinor, Kate Crookes, Linda Jeffery, and Daniel Dilks. 2012. A critical review of the development of face recognition: experience is less important than previously believed. *Cognitive Neuropsychology* 29 (1–2): 174–212.

Newton, Norman T. 1971. *Design on the Land: The Development of Landscape Architecture*. La Editorial, UPR.

Nightingale, Florence. 1969. *Notes on Nursing* (1860). New York, D. Appleton and Company.

Northwestern University Feinberg School of Medicine. 2014. "Scientists Discover Human Sperm Gene is 600 Million Years Old." Available at: www.feinberg.northwestern.edu/news/2010/2010F-July/Sperm_Gene.html (accessed 24 May 2014).

Placzek, Adolf K. 1965. *The Four Books of Architecture*. New York: Dover Publications.

Pinker, Steven. 2003. *The Blank Slate: The Modern Denial of Human Nature.* Penguin.

Platt, Rutherford H. 1996. *Land Use and Society: Geography, Law and Public Policy*. Washington, DC: Island Press.

Prosser, Wendy. 2012. Animal Body Plans and Movement: Symmetry in action. In *Decoded Science* [database online]. Available at: www.decodedscience.com/animal-body-plans-symmetry-in-action/13171 (accessed December 1, 2013).

Purves, Alexander. 1982. The persistence of formal patterns. *Perspecta* 19 : 138–63.

Ramachandran, V. S., and William Hirstein. 1999. The science of art: A neurological theory of aesthetic experience. *Journal of Consciousness Studies* 6 (6–7): 15–51.

Roth, G., and Dicke, U. 2005. Evolution of the brain and intelligence. *TRENDS in Cognitive Sciences* 9(5): 250–7.

Sagan, Carl. 1995. *The Demon-Haunted World: Science as a Candle in the Dark*. 1st ed. New York: Random House.

Schnörr, Stephanie J., Peter J. Steenbergen, Michael K. Richardson, and Danielle L. Champagne. 2012. Assessment of thigmotaxis in larval Zebrafish. In *Zebrafish Protocols for Neurobehavioral Research* 37–51. Springer.

Sea World Parks and Entertainment. 2014. "Elephants: Diet and Eating Habits." Available at: www.seaworld.org/animal-info/info-books/elephants/diet.htm (accessed February 9, 2014).

Snodgrass, J. J., Leonard, W. R., and Robertson, M. L. 2009. The energetics of encephalization in early hominids. In *The Evolution of Hominin Diets* Springer, Netherlands 15–29.

Sussman, Ann. 2011. More Vehicles Than People. In *Planetizen* [database online]. Available at: www.planetizen.com/node/48503 (accessed November 20, 2013).

Sussman, Ann. 2010. "Stories People Tell." *Open Neighborhood Project.* Available at: https://openneighborhood.blogspot.com (accessed May 24, 2014).

Taylor, R. P., B. Spehar, J. A. Wise, C. W. G. Clifford, B. R. Newell, C. M. Hagerhall, T. Purcell, and T. P. Martin. 2005. Perceptual and physiological responses to the visual complexity of fractal patterns. *Nonlinear Dynamics, Psychology, and Life Sciences* 9 (1): 89–114.

Taylor, R. P., B. Spehar, Van Donkelaar, P., and Hagerhall, C. M. 2011. Perceptual and Physiological Responses to Jackson Pollock's Fractals. *Frontiers in Human Neuroscience* 5: 60.

Thomas, Alan Ken (ed.) 2011. *The Business Wisdom of Steve Jobs: 250 quotes from the innovator who changed the world*. New York: Skyhorse Publishing, Inc.

Thompson, Peter. 1980. Margaret Thatcher: A New Illusion. *Perception* 9 (4): 483–4.

Ulrich, Roger S. 1984. View through a window may influence recovery from surgery. *Science* 224 (4647): 420–1.

Ulrich, Roger S. 2002. Health Benefits of Gardens in Hospitals. Paper presented at Plants for People Conference, Intl. Exhibition, Floriade.

Vartanian, Oshin, Gorka Navarrete, Anjan Chatterjee, Lars Brorson Fich, Helmut Leder, Cristián Modroño, Marcos Nadal, Nicolai Rostrup, and Martin Skov. 2013. Impact of contour on aesthetic judgments and approach-avoidance decisions in architecture. *Proceedings of the National Academy of Sciences* 110 (Supplement 2): 10446–53.

Vilotti, Jessica. 2013. Edgar Palmer's vision of Princeton, 75 years later. In *Aspire* [database online]. Available at: www.aspiremetro.com/edgar-palmers-vision-of-princeton-75-years-later/ (accessed December 1, 2013).

von Hoffman, Alexander. 2010. *Wrestling with Growth in Acton, Massachusetts: The possibilities and limits of Progressive Planning*. By Alexander von Hoffman Joint Center for Housing Studies, Harvard University, Cambridge, MA January 2010

Wald, Alan M. 1983. *The Revolutionary Imagination: The poetry and politics of John Wheelwright and Sherry Mangan*. Chapel Hill: UNC Press.

Weyl, Hermann. 1952. *Symmetry*. Princeton: Princeton University Press.

Wilson, Edward O. 1984. *Biophilia*. Harvard University Press.

Wilson, Edward O. 2006. *The Creation, An Appeal to Save Life on Earth*. New York City: W. W. Norton & Company.

Wilson, Edward O., and Stephen R. Kellert. 1993. *The Biophilia Hypothesis*. Washington DC: Island.

Wiseman, Carter. 2001. *I.M. Pei: A Profile in American Architecture*. Rev ed. New York: H.N. Abrams.

Wright, Frank Lloyd. 1907. A Fireproof House for $5,000. *Ladies Home Journal* 24.

Young, Kay. 2010. *Imagining Minds. The Neuro-Aesthetics of Austen, Eliot, and Hardy*. Ohio State University Press.

Young, Kay, and Jeffrey L. Saver. 2001. The Neurology of Narrative. *Substance* 72–84.

Further Reading

Ackerman, James. 1986. The Villa as Paradigm. *Perspecta* 22: 10–31.

Barnett, Jonathan. 2003. *Redesigning Cities: Principles, Practice, Implementation*. Chicago: Planners Press, American Planning Association.

Barrett, Maeve, Brendan Cullen, Corrina Maguinness, Niamh Merriman, Eugenie Roudaia, John Stapleton, Bernard MC Stienen, and Fiona N. Newell. 2012. A glance back on 50 years of research in perception. *The Irish Journal of Psychology* 33 (2–3): 65–71.

Barthes, Roland, and Lionel Duisit. 1975. An introduction to the structural analysis of narrative. *New Literary History* 6 (2): 237–72.

Benton, Michael J. 2008. *The History of Life*. New York: Oxford University Press Inc.

Besson, Morgane, and Jean-René Martin. 2005. Centrophobism/thigmotaxis, A new role for the mushroom bodies in drosophila. *Journal of Neurobiology* 62 (3): 386–96.

Biederman, Irving. 1987. Recognition-By-Components: A theory of human image understanding. *Psychological Review* 94 (2): 115.

Bilbo, Staci D., Lainy B. Day, and Walter Wilczynski. 2000. Anticholinergic effects in frogs in a Morris water maze analog. *Physiology and Behavior* 69 (3): 351–7.

Bloom, Nicholas Dagen. 2004. *Merchant of Illusion: James Rouse, America's Salesman of the Businessman's Utopia*. Urban Life and Urban Landscape Series. Columbus: Ohio State University Press.

Bratman, Gregory N., J. Paul Hamilton, and Gretchen C. Daily. 2012. The impacts of nature experience on human cognitive function and mental health. *Annals of the New York Academy of Sciences* 1249 (1): 118–36.

Brooks, Richard Oliver. *New Towns and Communal Values: A case study of Columbia, Maryland.* Special Studies in U.S. Economic, Social, and Political Issues. New York: Praeger.

Bruce, Vicki, and Andrew W. Young. 2012. *Face Perception*. London; New York: Psychology Press.

Burley, Jon, and Luis Loures. 2008. Conceptual Landscape Design Precedent: Four Historic Sites Revisited. *New Aspects of Landscape Architecture; Proceedings of 1st World Scientific and Engineering Academy and Society (WSEAS)*, Algarve, Portugal, WSEAS Press, 11–16.

Calthorpe, Peter, and William B. Fulton. 2001. *The Regional City: Planning For the End of Sprawl*. Washington, DC: Island Press.

Cárdenas, Rodrigo Andrés, and Lauren Julius Harris. 2006. Symmetrical decorations enhance the attractiveness of faces and abstract designs. *Evolution and Human Behavior* 27 (1): 1–18.

Cervero, Robert, and Peter Bosselmann. 1998. Transit Villages: Assessing the market potential through visual simulation. *Journal of Architectural and Planning Research* 181–96.

Conti, Flavio. 1978. *The Grand Tour: Shrines of Power.* Trans. Patrick Creagh. Boston: HBJ.

Dewitz, Johannes. 1886. Ueber gesetzmässigkeit in der ortsveränderung der spermatozoen und in der vereinigung derselben mit dem ei. I. *Pflügers Archiv European Journal of Physiology* 38 (1): 358–85.

Doolittle, John H. 1971. The effect of thigmotaxis on negative phototaxis in the earthworm. *Psychonomic Science* 22 (5): 311–2.

Duhigg, Charles. 2012. *The Power of Habit: Why We Do What We Do in Life and Business* 34. Random House Digital, Inc.

Dunham-Jones, Ellen, and June Williamson. 2009. *Retrofitting Suburbia: Urban design solutions for redesigning suburbs*. Hoboken, NJ: John Wiley & Sons.

Dunning, David, and Emily Balcetis. 2013. Wishful Seeing: How preferences shape visual perception. *Current Directions in Psychological Science* 22 (1): 33–7.

Editors of California Home and Design. 25 Buildings to Demolish Right Now. In *California Home+Design* [database online]. Available from www.californiahomedesign.com/inspiration/25-buildings-demolish-right-now/slide/5074.

Findlay, John M. 1992. *Magic Lands: Western Cityscapes and American Culture After 1940*. Berkeley: University of California Press.

Frantz, Douglas, and Catherine Collins. 2000. *Celebration, U.S.A.: Living in Disney's brave new town*. 1 Owl Books ed. New York: Henry Holt & Co.

Gazzaniga, Michael S. 1998. *The Mind's Past*. University of California Press.

George Washington's Mount Vernon. 2014. "Our Mission." Accessed February 9. www.mountvernon.org/about/our-mission.

Globe Staff. Boston City Hall Tops Ugliest-Building List in *The Boston Globe* [database online]. [cited November 14 2008]. Available from www.boston.com/news/local/breaking_news/2008/11/boston_city_hal_1.html (accessed December 1, 2013).

Goldsteen, Joel B., and Cecil D. Elliott. 1994. *Designing America: Creating urban identity: A primer on improving US cities for a changing future using the project approach to the design and financing of the spaces between buildings*. Van Nostrand Reinhold.

Goleman, Daniel. 1995. *Emotional Intelligence.* New York: Bantam Dell.

Gottdiener, Mark. 1997. *The Theming of America: Dreams, visions, and commercial spaces*. Boulder, Colorado: Westview Press.

Hagerhall, Caroline M., Thorbjörn Laike, Richard P. Taylor, Marianne Küller, Rikard Küller, and Theodore P. Martin. 2008. Investigations of Human EEG Response to Viewing Fractal Patterns. *Perception* 37 (10): 1488–94.

Hale, Jonathan. 1994. *The Old Way of Seeing*. New York: Houghton.

Han, Shihui, and Glyn W. Humphreys. 1999. Interactions between perceptual organization based on gestalt laws and those based on hierarchical processing. *Perception and Psychophysics* 61 (7): 1287–98.

Heath, Tom, Sandy G. Smith, and Bill Lim. 2000. Tall buildings and the urban skyline. The effect of visual complexity on preferences. *Environment and Behavior* 32 (4) (Jul 2000): 541–56.

Heller, Gregory L. 2013. *Ed Bacon: Planning, politics, and the building of modern Philadelphia*. University of Pennsylvania Press.

Hersey, George L. 1988. *The Lost Meaning of Classical Architecture.* Cambridge: MIT.

Herzog, Thomas R., and Jennifer A. Flynn-Smith. 2001. Preference and perceived danger as a function of the perceived curvature, length, and width of urban alleys. *Environment and Behavior* 33 (5): 653–66.

Hildebrandt, Andrea, Oliver Wilhelm, Grit Herzmann, and Werner Sommer. 2013. Face and object cognition across adult age. *Psychology and Aging* 28 (1) (March 2013): 243–8.

Historical Royal Palaces. 2014. "Hampton Court Maze." Available at www.hrp.org.uk/HamptonCourtPalace/maze (accessed February, 9).

Hochstein, Shaul, and Merav Ahissar. 2002. View From the Top: Hierarchies and reverse hierarchies in the visual system. *Neuron* 36 (5): 791–804.

Hoffman, James E. 1975. Hierarchical stages in the processing of visual information. *Perception and Psychophysics* 18 (5): 348–54.

Hummel, John E., and Irving Biederman. 1992. Dynamic binding in a neural network for shape recognition. *Psychological Review* 99 (3): 480.

Hummel, John E., and Brian J. Stankiewicz. 1998. Two Roles for Attention in Shape Perception: A structural description model of visual scrutiny. *Visual Cognition* 5 (1–2): 49–79.

Jackson, Kathy Merlock. 2011. *West Disneyland and Culture: Essays on the Parks and their Influence*, edited by Kathy Merlock Jackson and Mark I. West.

Jennings, H. S. 1897. Studies on reactions to stimuli in unicellular organisms. *Journal of Physiology* XXI: 258–322.

Jones, Benedict C., Lisa M. DeBruine, and Anthony C. Little. 2007. The role of symmetry in attraction to average faces. *Perception and Psychophysics* 69 (8) (November 2007): 1273–7.

Joye, Yannick. 2007. Architectural Lessons From Environmental Psychology: The case of biophilic architecture. *Review of General Psychology* 11 (4): 305–28.

Katz, Peter, Vincent Scully, and Todd W. Bressi. 1994. *The New Urbanism: Toward an architecture of community*. New York: McGraw-Hill.

Kim, Gwang-Won, Gwang-Woo Jeong, Tae-Hoon Kim, Han-Su Baek, Seok-Kyun Oh, Heoung-Keun Kang, Sam-Gyu Lee, Yoon Soo Kim, and Jin-Kyu Song. 2010. Functional neuroanatomy associated with natural and urban scenic views in the human brain: 3.0 T functional MR imaging. *Korean Journal of Radiology* 11 (5): 507–13.

Kim, Tae-Hoon, Gwang-Woo Jeong, Han-Su Baek, Gwang-Won Kim, Thirunavukkarasu Sundaram, Heoung-Keun Kang, Seung-Won Lee, Hyung-Joong Kim, and Jin-Kyu Song. 2010. Human brain activation in response to visual stimulation with rural and urban scenery pictures: A functional magnetic resonance imaging study. *Science of the Total Environment* 408 (12): 2600–7.

Knudsen, Brian, Richard Florida, Kevin Stolarick, and Gary Gates. 2008. Density and creativity in US regions. *Annals of the Association of American Geographers* 98 (2): 461–78.

Lamprea, MR, FP Cardenas, J. Setem, and S. Morato. 2008. Thigmotactic responses in an open-field. *Brazilian Journal of Medical and Biological Research* 41 (2): 135–40.

Lang, Jon. 1987. *Creating Architectural Theory*.

Larsen, Kristin. 2008. Research in Progress: The Radburn idea as an emergent concept: Henry Wright's regional city. *Planning Perspectives* 23 (3): 381–95.

Le Corbusier. 1933. *The Brilliant City*. Paris, France: The Onion Press.

Le Corbusier. 1967. *The Radiant City*. Orion Press.

LeGates, Richard T., and Frederic Stout. 2003. *The City Reader*. Routledge. New York.

Levinson, David M. 2003. The Next America Revisited. *Journal of Planning Education and Research* 22 (4): 329–44.

Lindal, Pall J., and Terry Hartig. 2013. Architectural variation, building height, and the restorative quality of urban residential streetscapes. *Journal of Environmental Psychology* 33: 26–36.

Lubyk, Danielle M., Brian Dupuis, Lucio Gutiérrez, and Marcia L. Spetch. 2012. Geometric Orientation by Humans: Angles weigh in, *Psychonomic Bulletin and Review* 19 (3) (June 2012): 436–42.

Lynch, Kevin, and Joint Center for Urban Studies. 1960. *The Image of the City*. MIT paperback series. Vol. MIT 11. Cambridge, MA: MIT Press.

Marling, Karal Ann, Donna R. Braden, and Henry Ford. 2005. *Behind the Magic: 50 years of Disneyland*. Dearborn, MI: The Henry Ford.

McCullough, Robert. 2012. *A Path for Kindred Spirits: The Friendship of Clarence Stein and Benton MacKaye*. Center for American Places at Columbia College.

Mikoleit, Anne, and Moritz Pürckhauer. 2011. *Urban Code: 100 Lessons for Understanding the City*. Cambridge, MA: MIT Press.

Mlodinow, Leonard. 2013. *Subliminal: How Your Unconscious Mind Rules your Behavior*. Random House Digital, Inc.

Nasr, Shahin, and Roger B. H. Tootell. 2012. A cardinal orientation bias in scene-selective visual cortex. *The Journal of Neuroscience: The Official Journal of the Society for Neuroscience* 32 (43) (October 24, 2012): 14921–6.

O' Craven, Kathleen M., and Nancy Kanwisher. 2000. Mental imagery of faces and places activates corresponding stimulus-specific brain regions. *Journal of Cognitive Neuroscience* 12 (6): 1013–23.

O'Gorman, James F. 1971. The Villa Lante in Rome: Some drawings and some observations. *The Burlington Magazine* 113 (816): 133–8.

Ostwald, Michael J., Laura E. Tate, Noam Shoval, Bob McKercher, Amit Birenboim, Erica Ng, John Van Hoesen, Steven Letendre, Joanna Williams, and Michael Batty. 2013. The fractal analysis of architecture: Calibrating the box-counting method using scaling coefficient and grid disposition variables. *Environment and Planning B: Planning and Design* 40: 644–63.

Palermo, Liana, Laura Piccardi, Raffaella Nori, Fiorella Giusberti, and Cecilia Guariglia. 2012. The roles of categorical and coordinate spatial relations in recognizing buildings. *Attention, Perception and Psychophysics* 74 (8): 1732–41.

Rakover, Sam S. 2013. Explaining the face-inversion effect: The face–scheme incompatibility (FSI) model. *Psychonomic Bulletin and Review* 20 (4): 665–92.

Rasmussen, Steen Eiler. 1958. *Experiencing Architecture*.

Rentschler, Ingo, Martin Jüttner, Alexander Unzicker, and Theodor Landis. 1999. Innate and learned components of human visual preference. *Current Biology* 9 (13): 665–71.

Rocheleau, Matt. Foes critique Martin Walsh's city hall sale plan, in *The Boston Globe* [database online]. [cited September 16 2013]. Available from www.bostonglobe.com/metro/massachusetts/2013/09/15/boston-mayoral-candidate-state-rep-martin-walsh-proposes-selling-relocating-city-hall/8ZfEYFEYxnSjxMeB6wimnO/story.html (accessed December 1, 2013).

Rosenbaum, R. Shayna, Marilyne Ziegler, Gordon Winocur, Cheryl L. Grady, and Morris Moscovitch. 2004. "I have often walked down this street before": FMRI studies on the hippocampus and other structures during mental navigation of an old environment. *Hippocampus* 14 (7): 826–35.

Schickel, Richard. 1968. *The Disney Version: The Life, Times, Art, and Commerce of Walt Disney*. New York: Simon and Schuster.

Selhub, Eva M., and Alan C. Logan. 2012. *Your Brain on Nature: The science of nature's influence on your health, happiness and vitality*. Wiley.com.

Southworth, Michael, and Eran Ben-Joseph. 2004. Reconsidering the Cul-de-sac. *Access Magazine* 23–33. Available at: http://escholarship.org/ uc/item/1qn0g780#page-1 (accessed June 17, 2014).

Spehar, Branka, and Richard P. Taylor. 2013. Fractals in Art and Nature: Why do we like them? Paper presented at IS&T/SPIE Electronic Imaging.

Steel, Piers. 2011. *The Procrastination Equation*. New York: HarperCollins.

Steinbauer, Martin J. 2009. Thigmotaxis maintains processions of late-instar caterpillars of Ochrogaster lunifer. *Physiological Entomology* 34 (4): 345–9.

Tech day 2013—The Brain as Mind—Rebecca Saxe Ph.D. '03. In *MIT Tech TV* [database online]. [cited September 20 2013]. Available from http://ttv.mit.edu/genres/32-science/videos/25846-tech-day-2013-the-brain-as-mind-rebecca-saxe-phd-03 (accessed December 6, 2013).

Ulrich, Roger S. 1993. Biophilia, biophobia, and natural landscapes. *The Biophilia Hypothesis* 73–137.

Valtchanov, Deltcho, and Colin Ellard. 2010. Physiological and affective responses to immersion in virtual reality: effects of nature and urban settings. *Journal of CyberTherapy and Rehabilitation* 3.4

van der Meer, Elke, Martin Brucks, Anna Husemann, Mathias Hofmann, Jasmin Honold, and Reinhard Beyer. 2011. Human Perception of Urban Environment and Consequences for its Design. *Perspectives in Urban Ecology*. 305–31 Springer.

van Koningsbruggen, Martijn G., Marius V. Peelen, and Paul E. Downing. 2013. A causal role for the extrastriate body area in detecting people in real-world scenes. *The Journal of Neuroscience: The Official Journal of the Society for Neuroscience* 33 (16) (April 17, 2013): 7003–10.

Venturi, Robert, and Christopher Curtis Mead. 1989. *The Architecture of Robert Venturi*. University of New Mexico Press.

Watson, Donald, Alan J. Plattus, and Robert G. Shibley. 2003. *Time-saver Standards for Urban Design*. McGraw-Hill New York.

Welsh, Jonathan. Why Cars Got Angry. The Wall Street Journal [database online]. online.wsj.com. Available from: http://online.wsj.com/news/articles/SB114195150869994250 (accessed December 6, 2013).

Whitaker, Craig. 1996. *Architecture and the American Dream*. New York: Clarkson N. Potter.

Williams, Katie. *Pursuing Prototypes: Defining the principles of various garden design styles*. Available at: http://lamar.colostate.edu/~larch/students/362paper_Katie-Williams2003.pdf (accessed June 17, 2014).

Xu, Mingdi, Johan Lauwereyns, and Keiji Iramina. 2013. An event-related potential study using repetition priming to investigate different stages in face and building processing. Paper presented at World Congress on Medical Physics and Biomedical Engineering. May 26–31, 2012. Beijing, China.

Xu, Yaoda. 2005. Revisiting the role of the fusiform face area in visual expertise. *Cerebral Cortex* 15 (8) (August 01): 1234–42.

Zukin, Sharon. 1991. *Landscapes of Power: From Detroit to Disney World*. Berkeley, CA: University of California Press.

Index

Page numbers in *italics* refer to illustrations or captions

Lightning Source UK Ltd.
Milton Keynes UK
UKHW010255291019
352497UK00020B/288/P